Mathematics for Computer Algebra

Maurice Mignotte

Mathematics for Computer Algebra

Translated by Catherine Mignotte

Springer-Verlag
New York Berlin Heidelberg London Paris
Tokyo Hong Kong Barcelona Budapest

Maurice Mignotte
Université Louis Pasteur
Départment de Mathématique
67084 Strasbourg
France

Mathematics Subject Classification: 11Y05, 11Y11, 12D50, 12Y05, 13P05, 68Q40

This book was originally published in French by the Presses Universitaires de France, 1989. The French edition is entitled *Mathématiques pour le calcul formel.*

Library of Congress Cataloging-in-Publication Data
Mignotte, Maurice.
[Mathématiques pour le calcul formel. English]
Mathematics for computer algebra / Maurice Mignotte : translated by Catherine Mignotte.
p. cm.
Translation of: Mathématiques pour le calcul formel.
Includes bibliographical references and index.
ISBN 0-387-97675-2
1. Algebra – Data processing. I. Title.
QA155.7.E4M5213 1991
512 – dc20 91-33024

Printed on acid-free paper.

Production managed by Henry Krell; manufacturing supervised by Robert Paella.
Camera ready copy provided by the author.
Printed and bound by R.R. Donnelley and Sons, Harrisonburg, VA.
Printed in the United States of America.

9 8 7 6 5 4 3 2 1

ISBN 0-387-97675-2 Springer-Verlag New York Berlin Heidelberg
ISBN 3-540-97675-2 Springer-Verlag Berlin Heidelberg New York

PREFACE

This book corresponds to a mathematical course given in 1986/87 at the University Louis Pasteur, Strasbourg.

This work is primarily intended for graduate students. The following are necessary prerequisites : a few standard definitions in set theory, the definition of rational integers, some elementary facts in Combinatorics (maybe only Newton's binomial formula), some theorems of Analysis at the level of high schools, and some elementary Algebra (basic results about groups, rings, fields and linear algebra).

An important place is given to exercises. These exercises are only rarely direct applications of the course. More often, they constitute complements to the text. Mostly, hints or references are given so that the reader should be able to find solutions.

Chapters one and two deal with elementary results of Number Theory, for example : the euclidean algorithm, the Chinese remainder theorem and Fermat's little theorem. These results are useful by themselves, but they also constitute a concrete introduction to some notions in abstract algebra (for example, euclidean rings, principal rings ...). Algorithms are given for arithmetical operations with long integers.

The rest of the book, chapters 3 through 7, deals with polynomials. We give general results on polynomials over arbitrary rings. Then polynomials with complex coefficients are studied in chapter 4, including many estimates on the complex roots of polynomials. Some of these estimates are very useful in the subsequent chapters.

Chapter 5 introduces polynomials with real coefficients. The main theme of this chapter is the separation of the real roots of real polynomials. We recall many results of the last century, which generally do not appear in modern textbooks. Among them are Sturm's method, the rules of Descartes and Budan-Fourier and Vincent's theorem. These works are very important in real geometry, a domain which is now very active.

The sixth chapter deals with polynomials over finite fields. It plays an essential rôle in this book. It contains Berlekamp's algorithm to factorize polynomials with coefficients in finite fields. Many exercises were inspired by the very complete book of R. Lidl and H. Niederreiter, *Finite fields* which was published in 1983 by Addison Wesley.

In the last chapter, we study methods to factorize polynomials with integer coefficients. We present the famous L^3 algorithm of Lenstra-Lenstra-Lováz. This algorithm uses many of the results of the preceding chapters.

This course has been much influenced by the remarkable book of D. Knuth *The Art of Programming*, vol. 2, second edition, Addison-Wesley.

First published in French under the title "Mathématiques pour le calcul formel", P.U.F., Paris, 1989, it was translated by my wife, Catherine Mignotte, to whom I am very grateful.

More than twenty exercises have been added to the French edition, as well as two sections about companions matrices and linear recursive sequences and an appendix about determinants.

The English version has been read by my colleague Philippe Glesser and by Professors Attila Pethö and David Saunders who corrected many mistakes and made valuable suggestions. I am very grateful to them.

This new version was typed using TEX, and I am very grateful to my colleague Raymond Séroul who helped me to use this system.

Maurice Mignotte
Strasbourg, September 1991

CONTENTS

NOTATIONS

$A[X, Y, \ldots]$	ring of polynomials over A in X, Y, ...	85
$\deg(P)$	degree of the polynomial P	86
$\deg_X(P)$	degree of P relatively to X	86
$(F, G, \ldots)$	ideal generated by F, G, ...	90
order_p	order of x relatively to p	97
F'	derivative of the polynomial F	102
$F^{(i)}$	derivative of order i of the polynomial F	102
$\mathrm{res}_X(F, G)$	resultant of F and G relatively to X	105
$\mathrm{Res}_X(F, G)$	resultant of F and G	105
$\mathrm{Discr}(F)$	discriminant of the polynomial F	107
$\mathrm{ord}(F, a)$	order of F at the point a	121
$\mathrm{M}(P)$	measure of the polynomial P	149
$\mathrm{M}(\alpha)$	measure of the algebraic number α	149
$\mathrm{L}(P)$	length of the polynomial P	149
$\mathrm{H}(P)$	height of the polynomial P	149
$\|\|P\|\|$ or $\|\|P\|\|_2$	quadratic norm of the polynomial P	151
$\mathrm{sep}(P)$	minimal distance between distinct roots of P	166
$\mathrm{V}(x)$	number of changes of sign ... (Sturm)	193
$\mathrm{v}(x)$	*Idem* (in the sense of Budan-Fourier)	196
$\mathbb{F}_q$	finite field of q elements	230
Ω_p	algebraic closure of $\mathbb{F}_p$	230
${\mathbb{F}_q}^*$	group of the non zero elements of $\mathbb{F}_q$	230
Φ_n	cyclotomic polynomial of order n	235
$\mu(n)$	Möbius function of n	236
$\omega(F)$	number of irreducible factors of F	239
$L : A \longrightarrow A$	Berlekamp's mapping $\alpha : \alpha^q \mapsto \alpha$	248
$\#E$	cardinality of the set E	277
$u^h \,\|\|\, v$	u^h di6des v, but u^{h+1} doesn't di6de v	285
$\det(L)$	determinant of the lattice L	302

Chapter 1

ELEMENTARY ARITHMETICS

This chapter deals with the representation of rational integers and describes the basic algorithms of arithmetics (addition, multiplication, division, ...). The study of theses examples introduces the fundamental notion of the cost of an algorithm.

1. Representation of an integer in base B

1. Lexicographical order

To represent the fundamental result on the representation of integers in some base, recall the definition of the *lexicographical order.*

Let $E_1, \ldots, E_r$ be ordered sets, we work on the set

$$E = E_1 \times \cdots \times E_r$$

and we consider the following relation R between the r-tuples $x = (x_1, \ldots, x_r)$ and $y = (y_1, \ldots, y_r)$:

$$x\,R\,y \quad \text{if} \quad \text{for the first index } i \text{ such that } x_i \neq y_i, \text{ we have } x_i < y_i.$$

It is easy to verify that R is a strict order relation. This order is called the lexicographical order on E (it is the classification order of the words in a dictionary).

The set E, with this order relation, is the *lexicographical product* of the sets E_i. When each E_i is totally ordered, then their lexicographical product is totally ordered too.

2. Development in base B : existence

The fundamental result of this paragraph is the following theorem.

THEOREM 1.1. — *Let B be an integer > 1. For k, a strictly positive integer, let $\mathbf{B}^k$ be the lexicographical product of k sets all equal to the set $\mathbf{B} = \{0, 1, \dots, B-1\}$. The application*

$$f_r : (r_0, r_1, \dots, r_{k-1}) \longmapsto \sum_{i=0}^{k-1} r_i\, B^{k-i-1}$$

is an isomorphism of the ordered set $\mathbf{B}^k$ on the set $\{0, 1, \dots, B^k - 1\}$.

Proof

We use induction on k. First of all, it is easy to verify that the image of f_k is contained in the set $\{0, 1, \dots B^k - 1\}$; then that the function f_k is strictly increasing. Hence the result ... □

For each rational integer $a > 0$, there is a smaller integer $k > 0$ such that $a < B^k$. According to Theorem 1.1, there is one and only one finite sequence $(r_0, \dots, r_{k-1})$ such that every r_i belongs to the set $\{0, 1, \dots, B-1\}$ and that a is given by the expression

$$a = \sum_{i=0}^{k-1} r_i\, B^{k-i-1}\,;$$

besides r_0 is positive (otherwise $a < B^{k-1}$).

The previous expression is called the *development* or *representation* of the rational integer a *in base B* — or *in scale B* — and it is written

$$a = (r_0, r_1, \dots, r_{k-1})$$

or $a = (r_0 r_1 \dots r_{k-1})_B$; the integers r_i are called the *digits* of a in base B.

Remark

The digits of the development of a in base B can, sometimes, be considered as symbols of an alphabet — alphabet $\mathbf{B}$ — and the second representation as a word on this alphabet, or as a list of numbers. The latter is used in computer algebra systems written in LISP.

Examples

• The most usual system is the decimal system in which the base B is equal to ten.

• Computers work with binary numbers, that is, in base two. The main reason for this is that development in base two uses only two digits (which correspond to two possible states of a unit cell). Moreover, elementary arithmetic operations are very simple in base two.

• To calculate with integers of an arbitrary size, it is convenient to work in a base B which has a value almost as big as the one allowed by the size of the word in the computer (for example, if the word contains sixteen bits, then B can be chosen with $B \leq 2^{15}$).

It is very useful to have the following estimation of an integer in term of its representation in base B.

PROPOSITION 1.1. — *If the positive rational integer n admits the following development in base B :*

$$n = \sum_{i=0}^{k} a_i \, B^i \,, \qquad \textit{with} \quad a_k \neq 0 \,,$$

then, we have

$$B^k \leq a_k B^k \leq n \leq B^{k+1} - 1 \,.$$

Proof

The only nontrivial inequality is the last one. Observe that

$$n \leq (B-1)\,(1 + B + \ldots + B^k) = B^{k+1} - 1 \,.$$

This ends the proof. □

3. From development to number

According to the previous theorem, the transformation is simply

$$f_r : (r_0, r_1, \ldots, r_{k-1}) \longmapsto \sum_{i=0}^{k-1} r_i \, B^{k-i-1} \,.$$

Thus, for example,

$$\begin{aligned}(1101110)_2 &= 2^6 + 2^5 + 2^3 + 2^2 + 2^1 \\ &= (64)_{\text{ten}} + (32)_{\text{ten}} + 8 + 4 + 2 \\ &= (110)_{\text{ten}} = 110 = \text{“one hundred and ten”}.\end{aligned}$$

It is also possible to represent this transformation with an *algorithm*, that is a computing process.

```
data : B, k, r_0, ..., r_{k-1} integers ≥ 0 ;
n := r_0 ;
for i := 1 to k − 1 do n := nB + r_i.
```

Development-number algorithm 1

4. From number to development in base B

The formula

$$a = \sum_{i=0}^{k-1} r_i\, B^{k-i-1}$$

leads to the relation

$$a = Bq + r_{k-1}, \quad \text{where} \quad 0 \le r_{k-1} < B;$$

thus r_{k-1} is the remainder of the euclidean division of the rational integer a by the number B,

$$r_{k-1} = a \bmod B \quad (\text{“}a \text{ modulo } B\text{”}).$$

Moreover the integer q, the quotient (also denoted by $q = a \text{ div } B$, as used in the programming language PASCAL), is given by the formula

$$q = \sum_{i=0}^{k-2} r_i\, B^{k-i-2}.$$

The next digit r_{k-2} is given by

$$r_{k-2} = q \bmod B,$$

and so on ...

This relation leads to a recursive procedure to compute the development of a positive rational integer a in a base B, which can be summarized by the following formula :

$$\text{dev}(a\,;\,B) = \big(\text{dev}(a \text{ div } B\,;\,B),\, a \bmod B\big).$$

This formula can be translated at once into a recursive definition of an algorithm for this development.

The associated iterative algorithm is as follows :

```
data : a integer > 0, B integer ≥ 2 ;
i := 0 ;
n := a ;
while n > 0 do
   begin
   |     x[i] := n mod B ;
   |     i := i + 1 ;
   |     n := n div B {div is the integer division}
   end.
```

Number-development algorithm 1

Example

Consider the integer $a = 67$ and the base $B = 3$. The computation of the development of a in base B can be represented by the following table.

i	n	x
0	67	1
1	22	1
2	7	1
3	2	2
4	0	

Which shows that $67 = (2111)_3$. Indeed,

$$2 \times 27 + 9 + 3 + 1 = 67.$$

The development in base B can also be found by computing the highest weight digits at first, that is r_0, then r_1, then r_2 ...

The second algorithm can be described as follows.

```
data : a integer > 0, B integer ≥ 2 ;
k := 0 ; x := 1 ; n := a ;
while a ≥ xB do
   begin
   |     k := k + 1 ;
   |     x := xB
   end ;
for i := 0 to k do
   begin
   |     r[i] := n div x ;
   |     n := n mod x ;
   |     x := x div B
   end.
```

Number-development algorithm 2

In the example of the search for the ternary development of number 67, at first we find $k = 3$, then the following table

i	n	x	r
0	67	27	2
1	13	9	1
2	4	3	1
3	1	1	1

which gives indeed $67 = (2111)_3$.

5. Case of general rational integers

The representation of any nonnegative rational integer n, only requires its sign and its absolute value : if $|n| = (a_m \dots a_0)_B$, we put

$$\mathrm{code}_B(n) = (\varepsilon, a_m, \dots, a_0),$$

where

$$\varepsilon = \begin{cases} +, & \text{if } n > 0, \\ -, & \text{otherwise.} \end{cases}$$

The sign function of an integer n is given by

$$\operatorname{sign}(n) = \begin{cases} +1, & \text{if } n > 0, \\ 0, & \text{if } n = 0, \\ -1, & \text{if } n < 0. \end{cases}$$

6. Comparing two numbers

Only two positive numbers will be considered, the general case can easily be inferred. This comparison test is a direct consequence of theorem 1. Let $a = (a_m \ldots a_0)_B$ and $b = (b_n \ldots b_0)_B$ be two positive numbers written in scale B, such that a_m and b_n are not zero.

We assume that the computer is able to compare two integers. Their absolute values must not exceed $2k$. The computer must also have a sign function, as well as a subtraction for these integers. The integers B, m and n are also supposed to be equal to or less than $2k$.

The following algorithm permits us to compare two positive integers a and b.

```
data  : a = (a_m ... a_0)_B, a_m ≠ 0, b = (b_n ... b_0)_B, b_n ≠ 0 ;
{the answer is r = sign(a − b)}
if m ≠ n
   then r := sign(m − n)
   else begin
        |    i := m ;
        |    while a_i = b_i do i := i − 1 ;
        |    if i = −1 then r := 0
        |                else r := sign(a_i − b_i)
        end.
```

Comparison between two positive integers

2. Addition

1. Case of two positive numbers

Let a and b be two positive integers coded in base B and let their respective lengths be m and n. Let us compute their sum c. Suppose

$$a = \sum_{i=0}^{m-1} a_i\, B^i \quad \text{and} \quad b = \sum_{i=0}^{n-1} b_i\, B^i.$$

For example $m \geq n$. Then

$$b = \sum_{i=0}^{n-1} b_i\, B^i \qquad \text{with} \quad b_n = \cdots = b_{m-1} = 0.$$

Thus,

$$c = \sum_{i=0}^{m-1} (a_i + b_i)\, B^i,$$

but it usually does not give the representation

$$c = \sum_{i=0}^{m-1} c_i\, B^i$$

of c in base b : don't forget the carries!

Let r be the carry at a given time, then the algorithm is the following.

```
data  : a = (a_{m-1} ... a_0)_B, a > 0, b = (b_{n-1} ... b_0)_B, b > 0 ;
r := 0 ;
k := m - 1 ;
for i := 0 to m - 1 do
  begin
  |    s := a_i + b_i + r ;
  |    if s < B then begin c_i := s ; r := 0 end
  |             else begin c_i := s - B ; r := 1 end  ;
  end  ;
if r > 0 then begin k := m ; c_m := 1 end.
```

Sum in base B

Exercise

Prove that this algorithm is correct.

Remark

The algorithm shows that the carry can only be 0 or 1; indeed we have always $a_i + b_i \leq 2B - 2$.

A natural choice for the cost of the former algorithm is the number of elementary operations (that is, operations on the integers that are smaller than B). This cost can easily be estimated, it is $\mathrm{O}\big(\max\{m, n\}\big)$.

Remark

This bound cannot be replaced by $\mathrm{O}\big(\min\{m, n\}\big)$ because of possible arbitrary long sequences of carries, as in

$$\sum_{i=0}^{m-1} (B-1)B^i + 1 = B^m.$$

Exercise

Let a and b be two positive rational numbers, coded in base B as above. Evaluate the probability for a sequence of consecutive carries of length k, with $n-1 \leq k \leq m-1$, to occur. Deduce a statistical study of the cost of the algorithm for the addition of two integers.

2. Case of two numbers of any sign

Let a and b be two integers coded in base B. We want to compute their sum c. We do not consider the trivial case in which at least one of the two numbers is zero.

If a and b are of the same sign then

$$c = \mathrm{sign}\,(a) \cdot \big(|a| + |b|\big),$$

and there is no problem.

If a and b have opposite signs, there are two cases :

- if $|a| \geq |b|$ then

$$c = \operatorname{sign}(a) \cdot \big(|a| - |b|\big),$$

- if $|a| < |b|$ then

$$c = \operatorname{sign}(b) \cdot \big(|b| - |a|\big).$$

In order to compute c we only have to know how to

1°) compare $|a|$ and $|b|$,

2°) compute the difference between two positive integers a' and b', with $a' > b'$.

Since we already know how to do the first operation we only have to study the second one. Again, we must consider the carries. The following algorithm realizes this operation.

```
data  : a = (a_m ... a_0)_B, b = (b_m ... b_0)_B, a > b > 0 ;
{we suppose a_m > 0 but only b_m ≥ 0, as above}
k := 0 ;
for j := 0 to m do
  begin
  |    s := a_j − b_j + k ;
  |    if s ≥ 0 then begin c_j := s ; k := 0 end
  |              else begin cj := s + B ; k := −1 end
  end.
```

Difference of two positive integers

Exercise

Prove that this algorithm is correct.

3. Subtraction

We have just studied how to add two integers of any sign; it should now be easy to make a subtraction with the formula

$$a - b = a + (-b).$$

4. Multiplication

We only consider the case of two positive integers. Compute $c = ab$. As in paragraph 2, consider again integers a and b given by the formulas

$$a = \sum_{i=0}^{m-1} a_i\, B^i \quad \text{and} \quad b = \sum_{i=0}^{n-1} b_i\, B^i ,$$

with $m \geq n$.

Again, we must get the development of c in base B. We have

$$c = \left(\sum_{i=0}^{m-1} a_i\, B^i\right) \cdot \left(\sum_{j=0}^{n-1} b_j B^j\right) = \sum_{h=0}^{m+n-2} \left(\sum_{i+j=h} a_i\, b_j\right) B^h .$$

Hence the algorithm (where r is the carry).

```
data  : a = (a_{m-1} ... a_0)_B, a_{m-1} ≠ 0,
        b = (b_{n-1} ... b_0)_B, b_{n-1} ≠ 0 ;
        r := 0 ; k := m + n - 2 ;
for h := 0 to m + n - 2 do
        begin
            t := r ;
            for i := max (0, h - n + 1) to min (h, m - 1) do
              t := t + a_i b_{h-i} ;
            c_h := t mod B ;
            r := t div B
        end ;
if r ≠ 0 then begin k := k + 1 ; c_k := r end.
```

Multiplication of two integers, version 1

To verify that this algorithm is correct, the only nonobvious fact is that c_k is exactly the left digit in the development of c in base B. To prove it, notice that the two inequalities $a < B^m$ and $b < B^n$ imply $c < B^{m+n}$; in other words, the integer k satisfies the inequality $k \leq m + n - 1$.

It is easy to verify that the number of operations occuring in this algorithm is at most $\mathrm{O}(mn)$.

However, this algorithm, written too quickly, has a serious drawback : intermediate integers occur with values that may be bigger than $(B-1)^2n$, where n is arbitrarilly large. These are not "elementary" operations : elementary operations deal with integers smaller than B. Therefore, this algorithm must be rewritten. In the following version, we take care of this difficulty.

```
data : a = (a_{m-1} ... a_0)_B, a_{m-1} ≠ 0,
           b = (b_{n-1} ... b_0)_B, b_{n-1} ≠ 0 ;
for i := 0 to m - 1 do c_i := 0 ;
for j := 0 to n - 1 do
        begin
            r := 0 ;
            for h := 0 to m - 1 do
               begin
                   t := a_h b_j + c_{h+j} + r ;
                   c_{h+j} := t mod B ;
                   r := t div B
               end ;
            c_{j+m} := r ;
        end.
```

Multiplication of two integers, version 2

Now the intermediate numbers t_k verify

$$t_k = a_h\, b_j + c_{h+j} + r_k \leq (B-1)^2 + (B-1) + r_k,$$

where $r_k = t_{k-1}$ div B (where $t_{-1} = 0$). This implies that they are not too large. More precisely, they satisfy the inequalities

$$t_0 \leq B(B-1), \quad r_1 < B,$$
$$t_1 \leq B(B-1) + (B-1) < B^2, \quad r_2 < B, \quad \ldots$$

Note that the numbers t_j are all strictly smaller than B^2 and that the r_js are all strictly smaller than B.

Remark

This algorithm is different from the usual algorithm. In the case of the "pupil" algorithm, one stores the intermediate products, and then one makes a global addition. Here, one computes partial additions. Drawbacks of the ordinary algorithm are the size of memory used is $m \times n$, and that intermediate computations are done on integers of order B.

Here is an example that shows the difference between the two methods (in both operations, the digits of the result are written in boldfaced characters).

$$
\begin{array}{cccccccc}
 & & & & 4 & 5 & 6 & 7 \\
 & & & \times & 6 & 7 & 8 & 9 \\
\hline
 & & & 4 & 1 & 1 & 0 & 3 \\
 & & 3 & 6 & 5 & 3 & 6 & \\
 & 3 & 1 & 9 & 6 & 9 & & \\
2 & 7 & 4 & 0 & 2 & & & \\
\hline
\mathbf{3} & \mathbf{1} & \mathbf{0} & \mathbf{0} & \mathbf{5} & \mathbf{3} & \mathbf{6} & \mathbf{3}
\end{array}
\qquad\qquad
\begin{array}{cccccccc}
 & & & & 4 & 5 & 6 & 7 \\
 & & & \times & 6 & 7 & 8 & 9 \\
\hline
 & & & 4 & 1 & 1 & 0 & \mathbf{3} \\
 & & 3 & 6 & 5 & 3 & 6 & \\
\hline
 & & 4 & 0 & 6 & 4 & \mathbf{6} & \ldots \\
 & 3 & 1 & 9 & 6 & 9 & & \\
\hline
 & 3 & 6 & 0 & 3 & \mathbf{3} & \ldots & \\
2 & 7 & 4 & 0 & 2 & & & \\
\hline
\mathbf{3} & \mathbf{1} & \mathbf{0} & \mathbf{0} & \mathbf{5} & \ldots & &
\end{array}
$$

5. Euclidean division

1. Existence

THEOREM 1.2. — *Let a and b be positive rational integers. Then there exists a unique pair of rational integers (q, r) such that $a = bq + r$, where $0 \leq r < b$. Set*

$$q = a \text{ div } b \qquad \textit{and} \qquad r = a \bmod b,$$

where the integer q is called the quotient *and r is the* rest.

When the rest of the division of a by b is zero one says that b divides *a, which is often written $b \mid a$; one also says that b is a* divisor *of a or that a is a* multiple *of b.*

Proof

The existence is proved by induction on a :

- if $a < b$, take $q = 0$ and $r = a$,
- if $a \geq b$, choose $q = 1 + \big((a-b) \text{ div } b\big)$ and $r = (a-b) \bmod b$.

The proof of unicity is left to the reader as exercise. $\square$

COROLLARY. — *Let H be a subgroup of $\mathbb{Z}$; then there exists a rational integer $d \geq 0$, and only one, such that $H = d\,\mathbb{Z}$ $(= \{dn\,;\, n \in \mathbb{Z}\})$.*

Proof

If $H = 0$, the result is true for $d = 0$. Suppose now that H contains non-zero elements. Then H contains a smallest positive element d. If a belongs to H, by euclidean division, $a = qd + r$, where $0 \leq r < d$. The formula $r = a - qd$ shows that r belongs to H. By the definition of d, one must have $r = 0$. Thus $a = qd$. $\square$

2. Computation

Let a and b be two positive integers of lengths m and n, respectively. We want to find two integers q and r such that $a = bq + r$, with $0 \leq r < b$.

The proof of the existence of the euclidean division gives a way to realize this operation, thanks to a sequence of subtractions. This method may be useful for large numbers, but first we study the following method.

Like in ordinary division, we may restrict ourselves to the case where $m = n + 1$. Then we put $a = (a_0\ a_1 \ldots a_n)_B$ and $b = (b_1 \ldots b_n)_B$ where $a/b < B$.

Next, we have to find the quotient $q = [a/b]$.

The principle of this method is to begin with an approximate value of the quotient q. Then the exact value is reached after a very small number of tries.

For this approximate value, we take

$$q^* = \min\left\{\left[\frac{a_0 B + a_1}{b_1}\right], B - 1\right\}.$$

The notation $[x]$ stands for the integer part of a real number x, the "floor" function, sometimes also denoted $\lfloor x \rfloor$. In others words, when x is a real number, $[x]$ is equal to the unique rational integer n, which satisfies the inequalities $n \leq x < n + 1$.

What follows shows that this first value is very satisfactory.

Indeed

(i) $\quad q^* \geq q$;

- this is true if $q^* = B - 1$,
- if $q^* \neq B - 1$, then

$$q^* = \left[\frac{a_0 B + a_1}{b_1}\right],$$

and $q^* b_1 \geq a_0 B + a_1 - b_1 + 1$, which implies

$$\begin{aligned} a - q^* b &\leq a - q^* b_1 B^{n-1} \\ &\leq a_0 B^n + \ldots + a_n - a_0 B_n - a_1 B^{n-1} + b_1 B^{n-1} - B^{n-1} \\ &= a_2 B^{n-2} + \ldots + a_n - B^{n-1} + b_1 B^{n-1} \\ &< b_1 B^{n-1} \leq b. \end{aligned}$$

The result follows, since q is the smallest integer such that $a - qb < b$.

(ii) $\quad b_1 \geq [B/2] \longrightarrow q^* \leq q + 2.$

Notice that

$$q^* \leq \frac{a}{b_1 B^{n-1}} < \frac{a}{b - B^{n-1}} \qquad \text{and} \qquad q > \frac{a}{b} - 1.$$

To prove this implication, one argues by *reductio ad adsurbum* : thus we may suppose that $q^* \geq q + 3$. Then

$$3 < \frac{a}{b - B^{n-1}} - \frac{a}{b} + 1 = \frac{a}{b}\left(\frac{B^{n-1}}{b - B^{n-1}}\right) + 1,$$

thus

$$\frac{a}{b} > 2\left(\frac{b - B^{n-1}}{B^{n-1}}\right) \geq 2(b_1 - 1),$$

and $B - 4 \geq q^* - 3 \geq q = [a/b] \geq 2(b_1 - 1)$. And we have $b_1 < [B/2]$.

Remark

The condition $b_1 \geq [B/2]$ is satisfied if a and b are respectively replaced by ka and kb, where k is given by the formula

$$k = \left[\frac{B}{b_1 + 1}\right].$$

Exercise

Prove the preceding remark.

6. The cost of multiplication and division

1. The cost of multiplication

Let $M(n)$ be the cost of the product of two integers of n bits. The algorithm of multiplication of two integers of length n given above has a cost of order n^2; thus $M(n) = \mathrm{O}(n^2)$. But it is possible to find algorithms for which the asymptotic cost is much smaller.

The following very easy method, due to Karatsuba †, uses an idea that is often very efficient in algorithmics : it splits the problem into two subproblems of equal size.

Suppose n is even; then $n = 2m$, and u and v are two integers of n bits. We write

$$u = a\,2^m + b, \quad v = c\,2^m + d,$$

and we compute the product $w = uv$ in the following way. If

$$x = (a+b)\,(c+d), \quad y = ac, \quad z = bd$$

then

$$w = y\,2^{2m} + (x - y - z)\,2^m + z.$$

† A.A. Karatsuba, in *Doklady Akad. SSSR*, 145, 1962, p. 293–294, in russian (english translation in *Soviet Physics-Doklady*, 7, 1967, p. 595–596.)

In the preceding formulas there are only three multiplications of integers of length m and extra operations (additions and shifts) whose cost is linear in n. There exists a constant k such that $M(n)$ is bounded above by

$$M(n) \leq \begin{cases} k & \text{if} \quad m = 1, \\ 3M(m) + kn & \text{if} \quad m > 1. \end{cases}$$

Let n be any positive integer and let t be the smallest positive integer such that $2^t \geq n$. Then $M(n) \leq M(2^t)$. Using the previous inequalities for the integers $2^t, 2^{t-1}, \ldots$ and 2, one gets

$$M(2^t) = \mathrm{O}(3^t),$$

which implies

$$M(n) = \mathrm{O}(n^\alpha) \quad \text{where} \quad \alpha = (\mathrm{Log}\ 3)/(\mathrm{Log}\ 2) \simeq 1.585 < 2.$$

This idea can be generalized. Instead of splitting the integers into two parts of same length, cut them into r pieces. Then, we have to find a way to compute the product using as few multiplications as possible. The technique is to obtain this product via a polynomial. This polynomial is determined by interpolation. Consider now two integers u and v of length equal to nr, we have to compute the product w.

Set

$$u = U_r 2^{rn} + \ldots + U_1 2^n + U_0$$

and

$$v = V_r 2^{rn} + \ldots + V_1 2^n + V_0 .$$

The polynomials U, V, and W are defined by the formulas

$$U(X) = U_r X^r + \ldots + U_1 X + U_0 ,$$

$$V(X) = V_r X^r + \ldots + V_1 X + V_0 ,$$

$$W(X) = U(X)V(X).$$

Then $w = W(2^n)$. To compute the polynomial W, whose degree is not bigger than $2r$, the first step is to compute its values at $2r+1$ points, for example $i = 0, 1, \ldots, 2r$ (or $-r, -r+1, \ldots, r-1, r$), thanks to the formula $W(i) = U(i)\,V(i)$. The coefficients of W are then obtained by Lagrange interpolation‡ (the cost of this operation is linear in n).

As a consequence, we get the inequality $M(rn) \le (2r+1)M(n) + cn$, for some positive constant c.

A proof similar to the previous one leads to the estimation

$$M(n) = \mathrm{O}(n^{\alpha'}) \quad \text{where} \quad \alpha' = \frac{\operatorname{Log}(2r+1)}{\operatorname{Log}(2r)}.$$

Taking r large enough, we obtain

$$M(n) = \mathrm{O}(n^{1+\varepsilon}),$$

for all fixed $\varepsilon > 0$.

Today the best known estimation is $M(n) = O(n \operatorname{Log} n \operatorname{Log}\operatorname{Log} n)$. This result is due to Schönhage and Strassen ‡, and it uses the technique of the fast Fourier transform. Since the interest in this result is mainly theoretical, we do not present it here.

2. The cost of division

Let us first consider the special case of the quotient of the integer 2^{2n-1} by an integer v, where v has n bits (thus $v \ge 2^{n-1}$). We exclude the trivial case $v = 2^n$. This quotient can be computed in the following way. Suppose that v is an integer, $2^{n-1} \le v < 2^n$. We have to compute

$$q = \left[2^{2n-1}/v\right].$$

We denote by $C(n)$ the cost of such a computation. To estimate $C(n)$, for convenience (and without loss of generality) we suppose the integer n to be even, say $n = 2m$.

‡ See Exercise 5, Chapter 3.

‡ A. Shönhage and V. Strassen, *Computing*, 7, 1971, p. 281–292.

We put $v = 2^m v' + v''$, so that $v' \geq 2^{m-1}$. Then

$$\frac{2^{2n-1}}{v} = \frac{2^{3m-1}}{v'}\,(1 - x + x^2 - \cdots),$$

where $x = v''/(2^m v')$; thus $0 \leq x < 2^{-m+1}$, which implies

$$0 \leq \frac{2^{2n-1}}{v} - \frac{2^{3m-1}}{v'}\,(1-x) < 4.$$

Now, we have to estimate the "error" term $(2^{3m-1}/v')\,(1-x)$. Set $2^{2m-1}/v' = a + \varepsilon$, where a is an integer and $0 \leq \varepsilon < 1$; then the error term is given by the following formula

$$(2^{3m-1}/v')\,(1-x) = 2^m a + 2^m \varepsilon - (2^{2m-1} v'')/v'^2,$$

where

$$\begin{aligned} 2^m \varepsilon &= (2^{2m-1} - av') \times (2^{2m-1}/v') \times 2^{-m+1} \\ &= (2^{2m-1} - av') \times a \times 2^{-m+1} + \varepsilon', \quad \text{with } 0 \leq \varepsilon' < 2. \end{aligned}$$

These formulas show that the quotient q can be obtained from the quotient a, the products $av', (2^{2m-1} - av') \times a$, and v'^2, the quotient $\left[(2^{2m-1}v'')/v'^2\right]$, and some extra simple operations. Moreover, the cost of these auxiliary operations is linear in n.

This leads to the inequality

$$C(n) \leq 2C(n/2) + 3M(n/2) + cn,$$

where c is some positive constant.

Under the hypothesis $M(2k) \geq 2M(k)$ for all k, this inequality implies the estimate

$$C(n) = \mathrm{O}\big(M(n)\big) + \mathrm{O}\,(n \operatorname{Log} n).$$

If one adds the extra hypothesis (which has yet not been proved or disproved)

$$M(n) \geq c' n \operatorname{Log} n,$$

for some positive constant c', the previous estimate leads to the interesting relation

$$C(n) = \mathrm{O}\big(M(n)\big).$$

To compute the quotient of a number u of $2n$ bits by a number v of n bits, just write

$$[u/v] = \left[(u/2^{2n-1}) \times (2^{2n-1}/v)\right].$$

Then the quotient is obtained by a computation of the preceding type, followed by a multiplication of two integers of approximately n bits.

If

$$D(n) := \text{ cost of the division of an integer of } 2n \text{ bits by one of } n \text{ bits},$$

we have proved that

$$D(n) = \mathrm{O}\big(M(n)\big).$$

One may say that the cost of division is dominated by the cost of multiplication, up to a multiplicative constant.

7. How to compute powers

Let n be a positive integer and x be any mathematical object such that the expression x^n has some sense (for example, x can be a rational integer, a matrix, an element of a finite field, an element of a monoid, and so on). One wants to compute x^n. More generally, we can consider any associative operation $\top$ and the computation of the term $x\top\cdots\top x$, n times.

1. First algorithm

This is the first algorithm one can imagine : compute successively the values x^2, x^3, ..., x^n.

```
data : x, n ;  {the result will be z}
z := x ;
for i = 1 to n − 1 do z := zx.
```

A1 : *computation of x^n, first version*

The natural cost of this algorithm is the number of multiplications. Here, the cost is Cost (A1) $= n - 1$.

Remark

The choice of the number of multiplications as a measure of the cost of this algorithm is not always very satisfactory. This question depends on the set in which we are computing. For example, if x is a rational integer, the time needed to compute the power x^n depends on the length of the integers that occur during the successive multiplications and this time is not at all linear with respect to n. On the contrary, if x belongs to a finite field, then the time for a multiplication may be considered as constant, and the total time of the computation of x^n is really proportional to the number of multiplications. For a detailed study of some special cases related to this remark, see the second algorithm.

2. Second algorithm

This algorithm has been known for approximately 2000 years by Chinese mathematicians and has been rediscovered many times. The principle is to compute successively the powers of x with an exponent that is a power of two (that is, the successive terms x^2, x^4, x^8, ...) and to combine these powers of x to obtain the term x^n, using the binary decomposition of the exponent n.

More precisely, if the representation of the exponent n in base two is given by the formula

$$n = \sum_{i=0}^{k} e_i \cdot 2^i , \quad e_i \in \{0, 1\},$$

then the term x^n can be written in the following ways

$$\begin{aligned} x^n &= x^{\sum_{i=0}^{k} e_i \cdot 2^i}, \\ &= \prod_{i=0}^{k} x^{e_i \cdot 2^i}, \\ &= \prod_{i;\, e_i \neq 0} x^{2^i}. \end{aligned}$$

This last formula leads almost at once to the following algorithm.

```
data : x, n ;
y := x ; z := 1 ; {the result will be z}
{the successive values of y are x, x^2, x^4, ...}
z := x ;
while n > 1 do
  begin
  |    m := m div 2 ;
  |    if n > 2m then z := zy ;
  |    y := yy ;
  |    n := m ;
  end ;
z := zy.
```

A2 : *computation of x^n, second version*

We run this algorithm for the computation of x^{13}. The following table gives the details of this computation.

n	13	13	6	3	1
m		6	3	1	
y	x	x^2	x^4	x^8	
z	1	x	x	x^5	x^{13}

Now, let us study the number N of multiplications made during this procedure. By the above formula

$$N = k + e_0 + e_1 + \ldots + e_k - 1 .$$

So that $N \leq 2k$.

We have to find an upper bound for the number k. The binary decomposition of n is

$$n = 2^k + e_{k-1} \cdot 2^{k-1} + \ldots + e_0 ;$$

thus $2^k \leq n$. Taking logarithms, we get the inequality $k \leq \log n$.

Remark

In the sequel, log will always be the binary logarithm, that is

$$\log x = \frac{\operatorname{Log} x}{\operatorname{Log} 2},$$

where Log is he usual neperian logatithm. Thus,

$$N \leq 2 \log k\,.$$

For example, if $n = 10^6$, algorithm A1 needs exactly 999, 999 multiplications, whereas algorithm A2 needs at most 38 (proof : $2^{10} = 1024$, thus we have $2^{20} > 10^6$).

Exercise

Compute N, when $n = 10^6$.

The above estimate of N shows that, for large enough values of n (say for $n \geq 20$) algorithm A2 is much cheaper that algorithm A1. But, to compare the actual performances of these two algorithms, it is important to indicate precisely the domain where the computations are made :

- If x is an integer of length h then very large numbers occur in the second algorithm. If we suppose that the cost of the product of two integers of respective lengths m and m' is approximatively cmm' (where c is a positive constant) then the cost of A2 satisfies the following estimate

$$C_2(n) \simeq c\,h^2 \sum_{i=0}^{k-1} 2^{2^{i+1}} + c\,h^2 \sum_{i=1}^{k} e_i \Big(\sum_{j=0}^{i-1} e_j\ 2^j\Big)\, 2^i\,,$$

where the e_i are binary digits of n (in particular $e_k = 1$), whereas the cost of the first algorithm is given by

$$C_1(n) = c\,h\,\big(h + 2h + 3h + \cdots + (n-2)h\big) \simeq \tfrac{1}{2}\,c n^2\,h^2\,.$$

For $n = 2^k$,

$$C_2(n) \simeq \tfrac{4}{3}\,c n^2\,h^2 \simeq \tfrac{8}{3}\,C_1(n)\,;$$

whereas, for $n = 2^k + 2^{k-1}$,

$$C_2(n) \simeq \tfrac{4}{3}\, c\, 2^{2k}\, h^2 + c\, 2^{2k-1}\, h^2 \simeq \tfrac{11}{6}\, c\, 2^{2k}\, h^2 ,$$

but

$$C_1(n) \simeq \tfrac{9}{4}\, c\ 2^{2k}\ h^2 \simeq \tfrac{27}{12}\, C_2(n) .$$

- If x belongs to a finite ring in which the cost of the product of two elements can be considered as constant, then the second algorithm is much faster than the first one for large enough n (for example for $n \geq 20$).

It is important to know that algorithm A2 is effectively used. In cryptography, for certain methods of encoding and of decoding, one has to compute expressions y such as

$$y = x^n \bmod a ,$$

where n is an integer bigger than 10^{50}. In such a case, the second algorithm is rather fast, but it is absolutely impossible to use the first one (exercise : estimate the time necessary for algorithm A1 to compute y on a computer which realizes 10^{10} multiplications per second, which is very optimistic!).

Now, come back to N for a more precise study. We saw that

$$N = k + \mathrm{s}(n) - 1 ,$$

where $\mathrm{s}(n)$ is the sum of the binary digits of n. In this formula, the integer k satisfies the inequalities $2^k \leq n < 2^{k+1}$, and thus $k \geq [\log n]$. Since $\mathrm{s}(n) \geq 1$, we have $N \geq [\log n]$, where equality occurs only for those values of n which are powers of two. To conclude,

$$[\log n] \leq N \leq 2\ [\log n] .$$

It would be interesting to do a statistical study of N, but this question is difficult and too specialized for us.

3. One application

The previous algorithm can be used to compute some term of a linear recursive sequence. Recall that such a sequence (u_n) is defined by its first values $u_0, u_1, \ldots, u_{h-1}$ and a condition like

$$(*) \qquad u_n = a_1 u_{n-1} + a_2 u_{n-2} + \ldots + a_h u_{n-h}, \quad \text{for} \quad n \geq h.$$

The most famous example is the Fibonacci sequence (which appears in 1202) and is defined by

$$F_0 = 0, \;\; F_1 = 1, \;\; F_n = F_{n-1} + F_{n-2}, \quad \text{for} \quad n \geq 2;$$

so that

$$F_2 = 1,\, F_3 = 2,\, F_4 = 3,\, F_5 = 5,\, F_6 = 8,\, F_7 = 13,\, F_8 = 21,\, F_9 = 34 \ldots$$

If U_n is the column vector, the components of which are $u_n, u_{n-1}, \ldots, u_{n-h+1}$ then relation $(*)$ is equivalent to the matrix formula

$$U_n = A\, U_{n-1},$$

where the matrix A is equal to

$$A = \begin{pmatrix} a_1 & a_2 & a_3 & \ldots & a_h \\ 1 & 0 & 0 & \ldots & 0 \\ 0 & 1 & 0 & \ldots & 0 \\ \ldots & \ldots & \ldots & \ldots & \ldots \\ 0 & 0 & \ldots & 1 & 0 \end{pmatrix}.$$

So that

$$U_n = A^{n-h+1}\, U_{h-1},$$

which shows that U_n, and in particular u_n, can be computed in $\mathrm{O}(\mathrm{Log}\, n)$ operations.

This method is especially useful when the computations are made modulo some integer M; for example, it gives a quick way to find the period of (u_n) modulo M.

4. Complexity of this problem

After the trivial algorithm to compute x^n, whose number of operations is linear in n, we gave another one that needs a number of operations logarithmic in n. Natural question : is it possible to find an even faster method? To answer such a question, we must give precise definitions.

First, define the class of algorithms to be considered. Here, we take the class of the algorithms that realize only multiplications. More precisely, for a fixed n, we consider only the algorithms to compute x^n of the following type :

$$1 \,:\, y_1 := u_1 v_1 \,, \quad \text{where} \quad u_1, v_1 \in \{1, x\},$$

$$\ldots\ldots$$

$$i \,:\, y_i := u_i v_i \,, \quad \text{where} \quad u_i, v_i \in \{1, x, y_1, \ldots, y_{i-1}\},$$

$$\ldots\ldots$$

$$t \,:\, y_t := u_t v_t \,, \quad \text{where} \quad u_t, v_t \in \{1, x, y_1, \ldots, y_{t-1}\} \text{ and } y_t = x^n .$$

The cost of such an algorithm is equal to t (the number of multiplications).

Remark that algorithm A2 belongs to this class. Notice also that this algorithm is not always optimal inside this class. Indeed, for $n = 15$, algorithm A2 needs six multiplications, whereas it is possible to compute x^{15} with only five operations, namely

compute x^3 (in two multiplications) then x^6, x^{12} and $x^{15} = x^3 \cdot x^{12}$. (It is possible to prove that algorithm A2 is optimal for $n < 15$.)

Now we study the *complexity* of the computation of x^n. In other words, we have to find a lower bound for the number t. For any algorithm of the preceding type, each value of y_i can be written

$$y_i := x^{e(i)} ,$$

where the exponents $e(i)$ satisfy the following conditions :

$$e(1) \in \{0, 1, 2\},$$

$$e(i) = e(j) + e(h) \quad \text{for some } \; j,\, h < i, \; \text{for } \; i \geq 2,$$

$$e(t) = n .$$

This implies the inequality

$$e(i) \leq 2^i \quad \text{for } i = 1, \ldots, t;$$

thus $e(t) = n \leq 2^t$. In other words,

$$t \geq \log n.$$

Now, we have proved that the complexity of this problem is at least $\log n$. This shows that algorithm A2 is almost the best possible among the class of procedures defined above.

Remark

The argument used here to find the complexity of this problem is very simple. This is very exceptional in the field of theoretical complexity. In that area most questions are very difficult. Most problems in complexity theory remain open. For example, the complexity of the product of two square matrices $n \times n$ is unknown for $n \geq 3$.

8. The g.c.d.

1. Existence. Relation of Bézout. Theorem of Euclid-Gauss

The following result proves the existence of the g.c.d.

THEOREM 1.3. — *Let a and b be two rational integers. There exists a positive rational integer d that divides both a and b and such that any common divisor of a and b divides also d. This integer d is called the g.c.d. of a and b and is denoted by* $\text{g.c.d.}(a, b)$, *or sometimes simply by* (a, b), *when no confusion is possible.*

Moreover, there exist two rational integers u and v such that

$$d = ua + vb.$$

Proof

Consider the set of rational integers given by $ma + nb$, where m and n run over $\mathbb{Z}$.

It is very easy to verify that this set is an additive subgroup of $\mathbb{Z}$. We already noticed that such a set is of the form $d\mathbb{Z}$, for a certain positive rational integer d. Moreover, by construction, there exist two integers u and v such that $d = ua + vb$.

To conclude, we only have to prove that d divides a and b and that every common divisor of a and b also divides d. Since

$$a = 1 \cdot a + 0 \cdot b,$$

the integer a belongs to the set $d\mathbb{Z}$, in other words d divides a; for the same reason, d divides b.

Consider now a rational integer c which divides both a and b. Then, there exist two integers a' and b' such that $a = ca'$ and $b = cb'$. Whence, the relation

$$d = ua + vb = (ua' + vb') \cdot c$$

which shows that c does divide d. □

When the g.c.d. of a and b is equal to 1, then a and b are said to be *relatively prime* or *coprime*. Theorem 1.3 has the following corollary.

THEOREM 1.4. — *Let a and b be two integers. Then a and b are relatively prime if, and only if, there are two integers u and v such that there is the relation (of Bézout, or — more exactly — of Bachet de Méziriac)*

$$ua + bv = 1.$$

Proof

According to theorem 1 the condition is necessary. This condition is also sufficient : if $ua + bv = 1$, then any integer that simultaneously divides a and b also divides 1, thus a and b are relatively prime. □

This theorem has the following corollary, sometimes called Euclid-Gauss theorem.

THEOREM 1.5. — *Let a and b be two relatively prime integers. If an integer c is divisible by both a and b, then it is also divisible by their product ab.*

Proof

Let u and v be two integers such that $ua + vb = 1$. Then

$$uac + vbc = c,$$

and since the product ab divides the numbers ac and bc, it also divides the integer c. ▯

2. How to compute the g.c.d.

The demonstration of Theorem 1.3 only proves the existence of the integer d but does not give any direct means for its computation. In practice, it is very useful to be able to compute the g.c.d. of two integers, which has been done for twenty two centuries with the famous algorithm of Euclid.

However, Euclid's algorithm only gives the value of the g.c.d, but it is often useful to know the value of the integers u and v of Theorem 1.3. Since they can be found rather easily, we directly show here the variant that also produces u and v.

Suppose $a > b > 0$; then, the algorithm of Euclid computes the following series of triples $W_i = (t_i, u_i, v_i)$ defined by the conditions

$$\begin{aligned} W_0 &= (a, 1, 0), \\ W_1 &= (b, 0, 1), \\ W_{i+1} &= W_{i-1} - q_{i+1} W_i, \qquad \text{where } q_{i+1} = [t_{i-1}/t_i] \text{ for } i \geq 2; \end{aligned}$$

stopping at the first integer j for which t_j is equal to zero. Then we claim that

$$d = t_{j-1} \quad \text{and} \quad d = u_{j-1}a + v_{j-1}b.$$

To demonstrate these last formulas, note that

$$t_i = u_i a + v_i b \quad \text{for} \quad i = 0, 1, 2, \ldots$$

and in particular

$$t_{j-1} = u_{j-1} a + v_{j-1} b,$$

thus $t_{j-1} \in a\mathbb{Z} + b\mathbb{Z}$.

We, now, have to demonstrate that t_{j-1} is indeed equal to d, the g.c.d. of the integers a and b. According to the demonstration of theorem 1.3, we have the relation $a\mathbb{Z} + b\mathbb{Z} = d\mathbb{Z}$, thus d divides the integer t_{j-1}.

Now, we only have to prove that t_{j-1} divides both a and b. But, by definition, t_{j-1} divides both t_{j-1} and t_{j-2} and the formula

$$t_{i-2} = t_i + q_i t_{i-1},$$

successively applied for the values $i = j-1,\ j-2,\ \ldots,\ 3,\ 2$ proves that the term t_{j-1} divides the integers $t_{j-3},\ t_{j-4},\ \ldots,\ t_1,\ t_0$. Since $t_1 = b$ and $t_0 = a$, we get the result. □

Example

The following table shows the computation of the g.c.d. of the integers 1812 and 1572.

i	0	1	2	3	4	5	6	7
t	1812	1572	240	132	108	24	12	0
u	1	0	1	−6	7	−13	59	
v	0	1	−1	7	−8	15	−68	
q			1	6	1	1	4	

Then we get g.c.d. $(1812, 1572) = 12$, and the relation

$$59 \times 1812 - 68 \times 1572 = 12.$$

3. The cost of the algorithm of Euclid

It is very interesting to analyze the cost of this algorithm. The most dramatic result is shown in theorem 1.6.

THEOREM 1.6 (G. Lamé, 1845). — *Let n be a positive integer. Let u be the smallest positive integer for which the algorithm of Euclid on the numbers u and v', needs n steps, for any positive integer v' smaller than u. And suppose that the algorithm of Euclid needs n steps for some positive integer v, with $v < u$. Then*

$$u = F_{n+1} \quad \textit{and} \quad v = F_n\,,$$

where F_k is the k-th number of Fibonacci.

Proof

Indeed, we make the successive divisions

$$u_0 = u = q_1 v + r_1\,,$$

$$u_1 = v = q_2 r_1 + r_2\,,$$

$$\cdots\cdots$$

$$u_{n-1} = q_n r_{n-1} + r_n = q_n r_{n-1}\,,$$

thus $u_{n-1} \geq 1$, $u_{n-2} \geq u_{n-1} + 1 \geq 2$, $u_{n-3} \geq u_{n-2} + u_{n-1}$, ... We find that $u_{n-j} \geq F_{j+1}$ for $j = 1, 2, \ldots, n$. The details are left to the reader. □

COROLLARY 1. — *If $0 \leq v \leq u$ are rational integers then the number n of steps of the algorithm of Euclid satisfies*

$$n \leq \frac{\operatorname{Log}(\sqrt{5}u)}{\operatorname{Log}(\frac{1+\sqrt{5}}{2})} - 1\,.$$

Proof

We leave the proof as an exercise. □

For a detailed statistical analysis of the number of necessary steps in the algorithm of Euclid, see the book of Knuth*.

* D.E. Knuth .— *The art of computer programming*, Addison-Wesley, Reading (Mass.), 1981 (second edition), § 4.5.3.

COROLLARY 2. — *If $0 \leq v < u$ are rational numbers, where u is a number of n bits. Then the cost of the algorithm of Euclid is* $\mathrm{O}(n^2)$.

Proof

It is an almost immediate consequence of corollary 1. □

4. The l.c.m.

If a and b are two nonzero integers, then by definition their l.c.m. is the integer

$$\text{l.c.m.}\,(a,b) = ab/(a,b)\,,$$

and sometimes we also denote the l.c.m. of a and b by $[a,b]$.

It is easy to verify that $[a,b]$ is a common multiple of the integers a and b and that any integer which is both a multiple of a and of b is also a multiple of $[a,b]$.

Exercise

Demonstrate the properties above.

Example

We saw above that the g.c.d. of the numbers 1812 and 1572 is equal to 12. This implies that

$$\text{l.c.m.}\,(1812, 1572) = (1812 \times 572)/12 = 237372\,.$$

9. The group $\mathrm{G}(n)$

1. Definition

Let $n \geq 2$ be an integer. By definition, $\mathrm{G}(n)$ is the set of the elements of the quotient ring $\mathbb{Z}/n\mathbb{Z}$ which are invertible. According to theorem 1.3, the set $\mathrm{G}(n)$ consists in the classes of integers prime to n. We write $\varphi(n)$ for the cardinality of the set $\mathrm{G}(n)$, it is the number of integers of the interval $\{1, 2, \ldots, n-1\}$ which are prime with n. This function is called Euler's phi function or Euler's totient function. Since the product of two invertible elements is also invertible, it is easy to see that $\mathrm{G}(n)$ is a multiplicative group.

Any integer ≥ 2 is said to be *prime* if it has no other divisor than itself and one. The prime numbers are 2, 3, 5, 7, 11, 13 ... It is known since Euclid that there exist infinitely many prime numbers (see Exercise 19).

From now on, the letter p will always designate a prime number. Any nonprime integer at least equal to two will be called *composite.*

If p is a prime number then all integers of the interval $\{1, 2, \ldots, p-1\}$ are coprime with p, thus

$$\varphi(p) = p - 1 .$$

For any integer $n \geq 2$, we even have the equivalence

$$n \text{ prime} \iff \varphi(n) = n - 1 .$$

2. The theorem of Euler

The following lemma is a very simple particular case of the theorem of Lagrange on finite groups.

LEMMA 1.1. — *Let G be a finite multiplicative commutative group of cardinality k. Then every element x of G satisfies*

$$x^k = 1 .$$

[The theorem of Lagrange is the same result, but for any finite group.]

Proof

Let x be any element of G. The function $y \mapsto xy$ of G onto itself is one to one. Then,

$$\prod_{y \in G} y = \prod_{y \in G} xy = x^k \prod_{y \in G} y ,$$

the last equality is due to commutativity. We get the result by simplifying by the term $\prod_{y \in G} y$. $\square$

The application of this lemma to the group $\mathrm{G}(n)$ gives the following result.

THEOREM 1.7 (Euler). — *Let a be an integer relatively prime to an integer $n \geq 2$; then*

$$a^{\varphi(n)} \equiv 1 \pmod{n}.$$

Example

Consider the case where $n = 56$ and $x = 19$. We verify that $\varphi(56) = 24$, and the computation modulo 56 follows :

$$\begin{aligned} x^2 &= 289 \equiv 9, \\ x^4 &\equiv 9^2 = 81 \equiv 25, \\ x^6 &\equiv 25 \times 9 = 225 \equiv 1, \end{aligned}$$

thus $x^{24} \equiv 1 \pmod{56}$, as asserted in the theorem.

3. Computation of the inverse in $\mathrm{G}(n)$

Let x be an element of the group $\mathrm{G}(n)$. We want to compute its inverse in this group $u = x^{-1}$. In other words, we must find an integer u, which satisfies the condition

$$ux \equiv 1 \pmod{n},$$

or two integers u and v, such that

$$ux + vn = 1.$$

Two methods are possible :

- use the algorithm of Euclid for the g.c.d. of x and n, which gives the integer u,
- use the theorem of Euler, which gives the solution

$$u = x^{\varphi(n)-1} \bmod n.$$

In both cases, the number of operations is $O(\text{Log}\, n)$, but the second method may be faster. [This is only true when $\varphi(n)$ is known. Nevertheless, there is no fast way of computing $\varphi(n)$; this is so true that the security of some cryptographic methods relies on the difficulty of computing the value $\varphi(n)$ when n is a very big integer.]

Remark

In this case, the cost of the computation of the inverse of an element is much higher then the cost of the computation of the product of two elements. In general, the costs of the operations in $G(n)$ and in $\mathbb{N}$ are very different (we have seen it for the computation of x^n).

4. Computation of the coefficients of the relation of Bézout

With the theorem of Euler it is possible to compute the coefficients of the relation of Bézout without using the algorithm of Euclid. Indeed, consider two positive integers a and b relatively prime. We must find two integers u and v such that $ua + vb = 1$. This relation implies the congruences

$$ua \equiv 1 \pmod b \quad \text{and} \quad vb \equiv 1 \pmod a.$$

Suppose that $\varphi(a)$ and $\varphi(b)$ are known. We get a solution (u, v) by taking

$$u = a^{\varphi(b)-1} \bmod b$$

and

$$v = \left(b^{\varphi(a)-1} \bmod a\right) - a.$$

According to the theorem of Euler, the number

$$ua + vb - 1$$

is divisible by a and by b, thus it is also divisible by the product ab (since a and b are relatively prime); we also easily verify that

$$|ua + vb - 1| < ab.$$

In fact, we get the relation $ua + bv = 1$.

Example

Take the values chosen before in the example of the algorithm of Euclid. We first write

$$1812 = 12 \times 151 \quad \text{and} \quad 1572 = 12 \times 131$$

and then

$$\varphi(151) = 150 \quad \text{and} \quad \varphi(131) = 130\,.$$

Verify that

$$151^{129} \bmod 131 = 59$$

and

$$(131^{149} \bmod 151) - 151 = -68\,,$$

thus we get again the relation

$$59 \times 1812 - 68 \times 1572 = 12\,.$$

Remark

For two integers a and b, we have the following equivalence

$$(a,b) = 1 \iff a^{\varphi(b)} \equiv 1 \pmod{b}\,.$$

We have already seen that the condition is necessary.

Conversely, if an integer d divides both a and b, it also divides the number $a^k \bmod b$ for any integer $k > 0$; thus

$$a^k \bmod b = 1 \implies d = 1\,.$$

When the number $\varphi(b)$ is known, the previous equivalence gives a fast way of testing whether a and b are relatively prime.

Exercise

Let n be an integer that is the product of two very large prime numbers p and q. Then demonstrate that it is easy to determine the factors p and q when Euler's function $\varphi(n)$ is known, and *vice-versa*.

10. The Chinese remainder theorem

1. The theorem

A special case of the following result was known by Chinese astronomers of antiquity.

THEOREM 1.8. — *Let m and n be two relatively prime integers, then for any couple of integers a and b the system of congruences*

$$\begin{cases} x \equiv a \pmod{m} \\ x \equiv b \pmod{n} \end{cases}$$

admits integers solutions; furthermore, the solution is unique up to a multiple of the product mn.

Proof

Consider the natural application

$$\mathbb{Z} \longrightarrow \mathbb{Z}/m\mathbb{Z} \times \mathbb{Z}/n\mathbb{Z}$$

which associates ($x \bmod m, x \bmod n$) to the integer x.

It is obvious that it is a morphism of rings. Its kernel is the set of integers that are both divisible by m and by n; thus, according to Theorem 1.3, it is the set of multiples of mn. Which leads to a one-to-one morphism

$$\mathbb{Z}/mn\mathbb{Z} \longrightarrow \mathbb{Z}/m\mathbb{Z} \times \mathbb{Z}/n\mathbb{Z}.$$

This injective application sends a finite set in a finite set with the same cardinality; hence this application is surjective. Hence the theorem. □

COROLLARY 1. — *Let $m_1, \ldots, m_k$ be integers that are relatively prime in pairs. Then, for any k-tuple of integers $(a_1, \ldots, a_k)$, the system of congruences*

$$x \equiv a_i \pmod{m_i},\ 1 \le i \le k,$$

admits an integer solution, moreover this solution is unique up to a multiple of the product $m_1 \cdots m_k$.

Proof

Proof by induction on k, using the previous theorem. □

COROLLARY 2. — *Let a and b be two relatively prime positive integers, then*

$$\varphi(ab) = \varphi(a) \cdot \varphi(b).$$

Thus, if $p_1, \ldots, p_r$ are different prime numbers and $k_1, \ldots, k_r$ are positive integers, then

$$\varphi(p_1^{k_1} \cdots p_r^{k_r}) = (p_1 - 1)\, p_1^{k_1 - 1} \cdots (p_r - 1)\, p_r^{k_r - 1}.$$

Proof

The second assertion is an immediate consequence of the first one; let us demonstrate the first one. According to the proof of Theorem 1.8, when a and b are two positive integers, there is a ring isomorphism

$$\mathbb{Z}/ab\mathbb{Z} \sim \mathbb{Z}/a\mathbb{Z} \times \mathbb{Z}/b\mathbb{Z}.$$

In this isomorphism, the invertible elements are sent on the invertible elements, hence

$$\mathrm{G}(ab) \sim \mathrm{G}(a) \times \mathrm{G}(b).$$

Considering the cardinality of each member, we get the relation. □

2. Constructive proof

The previous demonstration of the Chinese remainder theorem, unfortunately, does not give an algorithm for the computation of x (it only suggests a systematic search of this solution among the set of integers $0, 1, \ldots, mn - 1 \ldots$!).

Here is another proof which, this time, gives an algorithm that efficiently computes an integer solution x of the system

$$\begin{cases} x \equiv a \pmod{m} \\ x \equiv b \pmod{n} \end{cases}$$

considered in the Chinese remainder theorem.

Thus, let m and n be two relatively prime integers. The algorithm of Euclid gives two integers u and v, such that

$$um + vn = 1.$$

Then the integer

$$x = bum + avn$$

satisfies

$$x \equiv avn \equiv a \pmod{m}$$

and

$$x \equiv bvm \equiv b \pmod{n},$$

thus, it is a solution for the system.

Example

Solve the system

$$\begin{cases} x \equiv 13 \pmod{151}, \\ x \equiv 31 \pmod{131}. \end{cases}$$

We have seen that $59 \cdot 151 - 68 \cdot 131 = 1$. Thus we can take

$$x = (31 \times 59 \times 151 - 13 \times 68 \times 131) \bmod (151 \times 131) = 2127.$$

3. Effective computations

Let $m_1, \ldots, m_k$ be positive integers relatively prime in pairs, and let M be the product of these integers. We look for an algorithm to compute the isomorphism

$$\psi : \mathbb{Z}/M\mathbb{Z} \xrightarrow{\sim} \prod_{i=1}^{k} (\mathbb{Z}/m_i\mathbb{Z})$$

and its inverse ψ^{-1}.

The computation of ψ is immediate :

$$\psi(x) = (x \bmod m_1, \dots, x \bmod m_k).$$

We can compute the reciprocal application as follows. Put

$$M_i = M/m_i \quad (\text{thus } (M_i, m_i) = 1), \text{ for } i = 1, \dots, k.$$

Then we compute integers $e_1, \dots, e_k$, such that

$$e_i\, M_i \equiv 1 \pmod{m_i}, \text{ for } i = 1, \dots, k.$$

Then we instantly verify that

$$\psi^{-1}(a_1, \dots, a_k) = (a_1 e_1 M_1 + \dots + a_k e_k M_k) \bmod M.$$

As we have just seen it, the numbers e_i can be computed with the algorithm of Euclid or with the theorem of Euler.

11. The prime numbers

Recall that any integer ≥ 2 is said to be prime if its only positive divisors are one or itself. According to the definition of a prime number, it is obvious that

$$\varphi(p) = p - 1,$$

and, according to the previous paragraph, $\mathbb{Z}/p\mathbb{Z}$ is a field.

A fundamental consequence of theorem 1.4 is the following.

THEOREM 1.9 (Fermat). — *Let p be a prime number and a an integer not divisible by p, then*

$$a^{p-1} \equiv 1 \pmod{p}.$$

COROLLARY. — *Let p be a prime number and a any rational integer, then*

$$a^p \equiv a \pmod{p}.$$

Proof

Indeed, when p does not divide a, it is an obvious consequence of Theorem 1.9, otherwise the two sides of the above relation are equal to zero. □

Remark

The theorem of Fermat gives a necessary condition of primality; however, this condition is not sufficient. There are numbers, called *Carmichael numbers*†, that satisfy the conclusion of the theorem of Fermat but are not prime; the two smallest are $561 = 3 \times 11 \times 17$ and $1729 = 7 \times 13 \times 19$. It is not known whether there is an infinity of Carmichael numbers.

The following result is well known.

THEOREM 1.10. — *Any integer* ≥ 2 *is equal to a product of prime factors, moreover this factorization is unique — up to the order of the factors.*

Hence the following proposition.

COROLLARY. — *If* m *and* n *are two positive integers respectively of the form*

$$m = p_1^{a_1} \cdots p_r^{a_r} \quad \textit{and} \quad n = p_1^{b_1} \cdots p_r^{b_r}$$

where p_i *are prime numbers and* a_i *and* b_i *are nonnegative numbers, then the* g.c.d. *of the integers* m *and* n *is given by the formula*

$$\text{g.c.d.}\,(m,n) = p_1^{c_1} \cdots p_r^{c_r}\,,$$

where $c_i = \min\{a_i, b_i\}$, $1 \leq i \leq r$; *whereas the* l.c.m. *of these two integers is equal to*

$$\text{l.c.m.}\,(m,n) = p_1^{d_1} \cdots p_r^{d_r}\,,$$

where $d_i = \max\{a_i, b_i\}$, *for* $1 \leq i \leq r$.

† More precisely, the Carmichael numbers are those composite integers n, for which Fermat theorem holds for each a with $(a, n) = 1$. They were introduced by R.D. Carmichael in the paper "On composite numbers P which satisfy the Fermat congruence $a^{P-1} \equiv 1 \pmod{P}$", *Amer. Math. Monthly*, 19, 1912, p. 22–27.

Exercises

1. Let $b_1, b_2, \ldots$ be integers ≥ 2. For $n \geq 1$, we put $B_n = b_1 b_2 \cdots b_n$. Demonstrate that there is only one way of representing any nonnegative rational integer a by

$$a = a_0 + a_1 B_1 + a_2 B_2 + \ldots,$$

with the conditions $0 \leq a_i < b_{i+1}$ for all i. Find an algorithm to compute the integers a_i.

2. Let $a_1, a_2, \ldots, a_k, \ldots$ be strictly positive integers. Demonstrate that if the integers a_i verify the condition

$$a_i > a_1 + a_2 + \ldots + a_{i-1} \qquad \text{for } i = 2, 3, \ldots,$$

then there is at most one way of representing any positive integer n by

$$n = a_1 x_1 + a_2 x_2 + \ldots, \qquad \text{where } x_i \in \{0, 1\}.$$

When the sequence (a_i) satisfies this condition, give an algorithm which, with n given, either

- computes the x_i if n can be coded like above, or
- indicates that n is not of this form.

3. Let (F_n) be the series of the numbers of Fibonacci defined by the first values $F_2 = 1$, $F_3 = 2$, and the condition $F_n = F_{n-1} + F_{n-2}$ for $n \geq 4$. Prove that any positive integer a can be written

$$a = x_1 F_2 + x_2 F_3 + x_3 F_4 \ldots, \qquad \text{where } x_i \in \{0, 1\}.$$

Is this representation unique?

4. In the study of the algorithm of Karatsuba, we have neglected the fact that some intermediate integers could have a length bigger than $m = n/2$. Indeed the numbers $a+b$ and $c+d$ are of length equal to m. In order to compute their product through a procedure of multiplication of integers of size m, the computations must be decomposed. We write

$$a+b = \alpha\, 2^m + \beta, \quad c+d = \gamma\, 2^m + \delta, \qquad \text{where } \alpha, \gamma \in \{0,1\}.$$

Then

$$(a+b)\,(c+d) = \alpha\,\gamma\, 2^n + (\alpha\,\delta + \beta\,\delta)\, 2^m + \beta\delta.$$

This formula can be computed in four steps :

- product of the two bits α and γ,
- sum $\alpha\delta + \beta\gamma$ (let us recall that $\alpha, \gamma \in \{0,1\}$),
- products by some powers of two (= shifts),
- product of the two integers β and δ, of size m.

Prove that the given estimation of the cost of the algorithm is correct.

5. Let a and b be two nonnegative constants and k an integer, $k > 1$. We suppose that T is a nondecreasing function, $T : \mathbb{N} \longrightarrow \mathbb{R}$, for which

$$T(n) \le \begin{cases} b & \text{if } n = 1, \\ a\,T(n/k) + bn & \text{if } n = kn' \ (n' \in \mathbb{N}). \end{cases}$$

Demonstrate the estimates

$$T(n) \le \begin{cases} \mathrm{O}(n) & \text{if } a < k, \\ \mathrm{O}(n \operatorname{Log} n) & \text{if } a = k, \\ \mathrm{O}(n^{\operatorname{Log} a / \operatorname{Log} b}) & \text{if } a > k. \end{cases}$$

6. Let $T : \mathbb{N} \longrightarrow \mathbb{R}$, be a nondecreasing function with $T(1) = 1$. Study the asymptotic behavior of T in the following cases :

(i) $T(n) = a\,T(n-1) + b\,n^c$,

(ii) $T(n) = T(n/2) + b\,n^c$,

(iii) $T(n) = T(n/2) + b\,n\,(\operatorname{Log} n)^c$,

where a, b, and c are strictly positive constants.

7. Let $(F_n)_{n\geq 0}$ be the sequence of Fibonacci defined by the initial conditions $F_0 = 0$, $F_1 = 1$, and the recursive relation

$$F_n = F_{n-1} + F_{n-2} \quad \text{for } \ n \geq 2.$$

Prove the inequality

$$F_n \geq \alpha^{n-2} \ \text{ for } \ n > 0,$$

where $\alpha = (1+\sqrt{5})/2 = 1.61803\ldots$ is the "golden number".

8. Let $\ell(n)$ be the minimal number r such that there exists a sequence of positive integers $a_0, a_1, \ldots, a_r$ with the conditions $a_0 = 1$ and $a_n = r$ and, such that for $i = 1, \ldots, r$ there exist indices j and k for which $a_i = a_j + a_k$, with $k \leq j < i$.

1°) What is the link between $\ell(n)$ and the computation of x^n ?

2°) Let $\mathrm{s}(n) = \mathrm{s}_2(n)$ be the sum of the digits of the binary representation of n. Prove the inequalities

(i) $\ell(n) \leq \log n + \mathrm{s}(n) - 1,$

(ii) $\ell(mn) \leq \ell(m) + \ell(n).$

3°) Let a and b be two integers, $a > b \geq 0$. Demonstrate the two relations

$$\ell(2^a) = a \quad \text{and} \quad \ell(2^a + 2^b) = a + 1\,.$$

9. Let B and B' be two fixed integers ≥ 2. Prove that the cost $C(n)$ of the conversion

$$(a_1, \ldots, a_n)_B \longmapsto (a'_1, \ldots, a'_m)_{B'}$$

satisfies $C(n) = \mathrm{O}\big(M(n)\,\mathrm{Log}\, n\big)$.

10. Consider the following algorithm.

```
data : a, b integers, a > b > 0 ;
while b > 0 do
  begin
  |   r := min (a mod b, b − (a mod b)) ;
  |   a := b ;
  |   b := r ;
  end ;
d := a.
```

algorithm ?

1°) Demonstrate that the algorithm ends with $d = \text{g.c.d.}\,(a, b)$.

2°) Find an upper bound for the number steps in the loop.

3°) Modify this algorithm to compute two integers u and v which are solutions of Bézout relation $ua + vb = 1$.

11. Suppose that the rational integer a is relatively prime with the k rational integers $n_1, n_2, \ldots, n_k$. Then, demonstrate that the integer a is also relatively prime with the product $n_1 \cdots n_k$ of these integers.

12. Let a, b, c be three positive integers. Prove the formula

$$\big((a,b),c\big) = \big(a,(b,c)\big).$$

13. If a rational integer a divides a rational integer b, then we write $a \mid b$. Let a, b, c be three positive integers. Prove that

$$c \mid ab \implies c \mid (a,c)\cdot(b,c).$$

14. Let x, y, z be three real numbers ≥ 0. Demonstrate the relation

$$\max\{x, y, z\} =$$
$$x + y + z - \min\{x,y\} - \min\{y,z\} - \min\{z,x\} + \min\{x,y,z\}.$$

Deduce the formula

$$[a,b,c] = \frac{abc}{(bc, ac, ab)},$$

15. Study the equation

$$ax \equiv b \pmod{m}$$

in the case of an integer $m \geq 2$.

16. Study the integer equation

$$ax + by = c.$$

17. Let a, b, c, d, e, f, and m be given integers, $m > 0$. Study the system

$$\begin{cases} ax + by \equiv e \pmod{m}, \\ cx + dy \equiv f \pmod{m}. \end{cases}$$

Apply your method to solve the system

$$\begin{cases} 2x + 3y \equiv 1 \pmod{21}, \\ 3x + 2y \equiv 5 \pmod{21}. \end{cases}$$

18. Consider the Chinese remainder theorem.

a) Demonstrate that the integers e_i satisfy

$$e_i^2 \equiv 1 \pmod{M} \quad \text{for } 1 \leq i \leq k,$$

and

$$e_i\, e_j \equiv 0 \pmod{M} \quad \text{for } 1 \leq i < j \leq k.$$

(In other words, the e_i are orthogonal idempotents of the ring $\mathbb{Z}/M\mathbb{Z}$.)

b) Let A be an integer which satisfies

$$A \equiv a_i \pmod{m_i} \quad \text{for } 1 \leq i \leq k.$$

Prove that there exists an integer E such that

$$\frac{A}{M} = E + \sum_{i=1}^{k} \frac{a_i}{m_i}.$$

It is the *standard decomposition* of the fraction A/M. Moreover, prove that decomposition is unique if we suppose that the integers a_i satisfy the condition $0 \leq a_i < m_i$ for $1 \leq i \leq k$.

19. a) Prove that there exist infinitely many prime numbers.

[Proof of Euclid : consider the integer $1 + 2 \cdot 3 \cdots p$ and notice that this integer is not divisible by any prime number $\leq p$.]

b) Demonstrate that there exists an infinity of prime numbers of the form $4n + 3$.

20. Consider the following algorithm, which computes the inverse of a binary fraction.

```
data : n, v = (0.v1 v2 ... vn)_2, v1 = 1, vi ∈ {0,1} ;
z := [32/(4v1 + 2v2 + v3)]/4 ;
k := 0 ; h := 1 ;
while h < n do
  begin
     h := 2h ; { here h = 2^(k+1) }
     V_k := (0.v1 v2 ... v_(h+3))_2 ;
     z := 2z − V_k z^2 ; { exact computation! }
     z := [2^(h+1) z + 0.5] 2^(−h−1) ;
     k := k + 1 ;
  end.
```

Computation of an inverse by the method of Newton

Prove that

(i) $0 < z \leq 2$,

(ii) $|z - 1/v| \leq 2^{-h}$ with $h = 2^k$, after k steps.

Deduce that the cost $C(n)$ of this algorithm is

$$2\,M(4\,n) + 2\,M(2\,n) + \cdots + O\,(n)$$

and thus that the usual hypothesis, $M(2\,m) \geq 2\,M(m)$ for all m, implies

$$C(n) \leq 4\,M(4\,n) + O\,(n)\,.$$

21. Let a and n be two integers ≥ 2. Demonstrate that n divides the number $\phi(a^n - 1)$. [Apply the theorem of Lagrange.]

22. Let p be prime number. Prove that

$$\binom{p-1}{j} \equiv (-1)^j \pmod{p},\ 0 \le j \le p-1.$$

Deduce that if a and b are two elements of a ring such that

$$p\,a = p\,b = 0 \quad \text{and} \quad a\,b = b\,a$$

then

$$(a-b)^{p-1} = \sum_{j=0}^{p-1} a^j\, b^{p-1-j}.$$

23. Let p be a prime number, k and m two integers verifying $1 \le k < p^m$. Let p^r be the highest power of p which divides k. Demonstrate that p^{m-r} is the highest power of p, which divides the binomial coefficient

$$\binom{p^m}{k}.$$

24. Compare in details the costs $C_1(n)$ and $C_2(n)$ defined in Section 7.2.

25. Let B be a rational integer, $B \ge 2$.

1°) Let $\mathrm{s}_B(x)$ be the sum of the digits of a positive integer x written in base B. Prove the inequality

$$\mathrm{s}_B(x+y) \le \mathrm{s}_B(x) + \mathrm{s}_B(y).$$

[First, suppose that y is equal to a power of B, and look at the possible carries. Then, consider the general case.]

2°) For any positive integer n, prove the relation

$$\sum_{k \ge 1} \left[\frac{n}{B^k}\right] = \frac{n - \mathrm{s}_B(n)}{B-1}.$$

26. Prove that each coefficient in the expansion of $(1-X^p)^{-1}-(1-X)^{-p}$ is divisible by p.

[Use the relation

$$(1-X^p)^{-1}-(1-X)^{-p}=(1-X)^{-p}\,(1-X)^{-1}\left((1-X)^p-1+X^p\right).]$$

27. Let α, β, $\gamma\ldots$ be properties relative to the elements of a set E of N elements. If π is any of the properties α, β, $\ldots$, $\alpha\beta$, $\alpha\gamma\ldots$ (where we have put $\alpha\beta=\alpha$ and $\beta,\ldots$), N_π designates the number of elements of E that satisfy property π.

1°) Prove that the number of elements of E that have none of the properties α, β, $\ldots$ is

$$N-N_\alpha-N_\beta-\cdots+N_{\alpha\beta}+\cdots-N_{\alpha\beta\gamma}-\cdots$$

[Suppose that an element x of E satisfies exactly k of the considered properties α, β, $\ldots$ Prove that the contribution of x in this sum is 1 if k is zero, but $\sum_{j=0}^{k}(-1)^j\binom{k}{j}=0$ if k is positive.]

2°) Deduce that the number of integers of the interval $\{1,\ldots,n\}$ which are not divisible by any one of a set of coprime integers a, b, $\ldots$ is equal to

$$[n]-\sum\left[\frac{n}{a}\right]+\sum\left[\frac{n}{ab}\right]-\cdots$$

As a special case, prove the formula $\varphi(n)=n\prod_{p|n}\left(1-\frac{1}{p}\right)$.

28. Cost of the euclidean algorithm

Let a and b be two positive integers, $a>b$. The aim of this exercise is to estimate the cost, say $E(a,b)$, of the computation of the g.c.d. of a and b, using the euclidean algorithm.

1°) Suppose that a and b have respectively m and n binary digits, with $m\geq n$. Demonstrate that the cost $D(a,b)$ of the euclidean division of a by b satisfies the relation

$$D(a,b)=\mathrm{O}\left(n\,(m-n+1)\right).$$

2°) If a and b are two positive integers such that $a > b$. Prove the inequality $a \bmod b \leq (a-1)/2$.

3°) Recall that the euclidean algorithm for a and b consists of computing the (decreasing) sequence of nonnegative integers $a_0 = a$, $a_1 = b$, $a_2 = a \bmod b$, ..., $a_{i+2} = a_i \bmod a_{i+1}$, ... until the number zero is reached; then the gcd of a and b is the last nonzero term of this sequence [say $(a,b) = a_k$].

(i) Using the second question, prove that the number k of steps of the euclidean algorithm on a and b satisfies $k \leq 2\log b + 2$ (recall that we put $\log x = \operatorname{Log} x / \operatorname{Log} 2$).

(ii) For $1 \leq i \leq k$, let n_i be the number of binary digits of a_i. Show that

$$E(a,b) \leq c \sum_{i=1}^{k-1} n_{i+1}(n_i - n_{i+1} + 1).$$

Conclude that

$$E(a,b) = \mathrm{O}\left(\operatorname{Log}(a) \operatorname{Log}(b)\right).$$

29. Let $(a_1, a_2, \dots, a_n)$ be a nonzero vector of n rational integers.

1°) Consider the following procedure : Let k be the number of non zero numbers among a_1, a_2, ..., a_n. While $k > 1$, do the following :

(i) Sort the a_is by decreasing order of absolute values, and let $(b_1, b_2, \dots, b_n)$ be the result.

(ii) Let i be the last index for which the integer b_i is not zero. Set $a_j = b_j$ for $j = i, i+1, \dots, n$ and replace each b_j, $2 \leq j < i$, by

$$a_j = \begin{cases} b_j \bmod |b_i|, & \text{if } (b_j \bmod |b_i|) \leq |b_i|/2, \\ \\ (b_j \bmod |b_i|) - b_i, & \text{otherwise.} \end{cases}$$

Suppose that the input of this procedure satisfies $|a_i| \leq A$, for $i = 1, 2, \dots, n$. Prove that this procedure stops after at most $\mathrm{O}(\operatorname{Log} A)$ steps and that the result is the vector $(\delta, 0, \dots, 0)$, where δ is the g.c.d. of the initial a_is. Moreover, show that each step of this procedure is equivalent to a matrix multiplication $V \mapsto VU$ where U is a *unimodular matrix* (that is, a matrix with determinant $= \pm 1$), and V is the vector $(a_1, a_2, \ldots, a_n)$. Conclude that there exists a unimodular matrix E such that

$$(a_1, a_2, \ldots, a_n)\, E = (\delta, 0, \ldots, 0).$$

2°) Let V be a nonzero vector of rational integers. Prove that there exists a matrix of rational integers whose first row is equal to V and whose determinant is equal to the g.c.d. of the components of this vector.

3°) Hermite form of a matrix

(i) Let A be a square matrix of size $n \times n$ over the integers. Prove that there exists a unimodular matrix U such that $T = UA$ is a lower triangular matrix, with nonnegative elements along the diagonal.

[Use induction on n. The result is trivial for $n = 1$. If $n \geq 2$, let ${}^t(\alpha_1, \alpha_2, \ldots, \alpha_n)$ be the last column of A. Let δ be the g.c.d. of the integers α_1, α_2, …, α_n (with the convention that that $\delta = 0$, if the α_is are all zero). Using question 1°, prove that there exists an invertible matrix E such that

$$E\,{}^t(\alpha_1, \alpha_2, \ldots, \alpha_n) = {}^t(0, \ldots, 0, \delta).$$

Now, it is easy to conclude.]

(ii) Moreover, show that using only elementary operations on rows, we can assume that the following property holds :

(*) all the elements t_{ij} of T belong to the interval $\{0, 1, \ldots, t_{ii} - 1\}$ for $i < j \leq n$. Such a matrix T is called a *Hermite form* of the matrix A.

(iii) Suppose that U_1 and U_2 are unimodular matrices such that $T_1 = U_1 A$ and $T_2 = U_2 A$ are Hermite forms for A. Demonstrate that T_1 and T_2 are equal (unicity of the Hermite form of a matrix).

[The hypothesis implies that $T_2 = UT_1$ where U is a lower triangular unimodular matrix. First show that the diagonal elements of U are all equal to 1. Then, if $U = (u_{i,j})$, $T_1 = (h_{i,j})$, $T_2 = (k_{i,j})$, using the relations

$$h_{s,s-1} = u_{s,s-1} k_{s-1,s-1} + k_{s,s-1}\,,$$

and property $(*)$, prove that $u_{s,s-1} = 0$ for $1 < s \le n$. More generally, using induction on the integer j, prove that $u_{s,s-j} = 0$ for all s, $j < s \le n$, and $j = 1, 2, \ldots, n-1$. Conclude that U is the identity matrix. Hence $T_1 = T_2$.

(iv) Show that

$$\mathrm{Herm}\begin{pmatrix} 2 & 5 & 2 \\ -7 & 3 & 6 \\ 1 & 1 & 4 \end{pmatrix} = \begin{pmatrix} 27 & 0 & 0 \\ 10 & 3 & 0 \\ -8 & -2 & 2 \end{pmatrix},$$

where $\mathrm{Herm}(A)$ denotes the Hermite form of a matrix A.

[Reference : Morris Newman .— *Integral matrices*, Academic Press, New York, 1972, chap. 2]

Chapter 2

NUMBER THEORY, COMPLEMENTS

This chapter contains some results of elementary number theory, which will be useful later and in some applications not considered here, such as cryptography.

1. Study of the group $\mathrm{G}(n)$

1. Some lemmas on finite groups

If x belongs to a group G, the order of x, denoted $\mathrm{order}_G(x)$, or simply $\mathrm{order}(x)$, is the smallest positive integer k, such that $x^k = 1$, if such a k exists, otherwise $\mathrm{order}_G(x) = \infty$. The order has the following properties.

PROPOSITION 2.1. — *Let x and y be two elements of a group G which commute and are of finite orders respectively equal to a and b. Then*

(i) $$\mathrm{order}_G(xy) \quad \textit{divides} \quad \mathrm{l.c.m.}(a,b)\,,$$

(ii) $$(a,b) = 1 \implies \mathrm{order}_G(xy) = ab\,.$$

Proof

(i) Let m be the l.c.m. of a and b; then

$$(xy)^m = x^m y^m = 1\,.$$

(ii) Suppose that the integers a and b are relatively prime and let u and v be two integers such that

$$ua + vb = 1\,.$$

Then, the relation

$$(xy)^{ua} = x^{ua}\, y^{1-vb} = y$$

shows that b divides $\mathrm{order}_G(xy)$. In the same way, a divides $\mathrm{order}_G(xy)$.

Since the integers a and b are relatively prime, their product ab divides $\mathrm{order}_G(xy)$. By relation (i), $\mathrm{order}_G(xy)$ divides the product ab. Hence the conclusion. ▯

PROPOSITION 2.2. — *Let G be a finite commutative group. Then G contains an element whose order is a multiple of* $\mathrm{order}(y)$, *for all y in the group G.*

Proof

Let x be an element of G whose order is maximal and let y be any element of G.

Suppose that the order of y does not divide the order of x. Then, there exists a prime number p and a positive integer h such that p^h divides $\mathrm{order}(y)$ but does not divide $\mathrm{order}(x)$. Put

$$\mathrm{order}(x) = p^{h'} a, \quad \text{with } 0 \le h' < h, \text{ and } (a,p) = 1,$$

$$\mathrm{order}(y) = p^h b,$$

$$x' = x^{p^{h'}} \quad \text{and} \quad y' = y^b.$$

Then

$$\mathrm{order}(x') = a, \quad \mathrm{order}(y') = p^h, \quad \text{where} \quad (a, p^h) = 1,$$

and, by Proposition 2.1, we get a contradiction :

$$\mathrm{order}(x'y') = p^h a > \mathrm{order}(x).$$

Hence the result. ▯

PROPOSITION 2.3. — *Let G be a finite group with n elements. Then the following properties are equivalent :*

(i) *G is cyclic,*

(ii) *for every divisor d of n, G contains at most d elements whose order divides d,*

(iii) *for every divisor d of n, G contains at most $\varphi(d)$ elements of order d.*

Proof

(i) $\Longrightarrow$ (ii) : Suppose G cyclic, say generated by x. Let d be a divisor of n. Then the elements of G whose order divides d are

$$1,\ x^m,\ x^{2m},\ \ldots,\ x^{(d-1)m} \quad \text{where} \quad m = n/d.$$

Hence the result.

(ii) $\Longrightarrow$ (iii) : Let d be a divisor of n. If G contains no element of order d then assertion (iii) is true. Otherwise, suppose that x is an element of G of order equal to d. Then the elements $1,\ x,\ x^2,\ \ldots,\ x^{d-1}$ are all distinct and the order of each of them is a divisor of d. If property (ii) holds, then these elements are the only elements of G whose order divides d. Thus, the only elements of G, the order of which is d, are the x^k, with $1 \leq k < d$ and $(k, d) = 1$; the number of these elements is exactly $\varphi(d)$.

(iii) $\Longrightarrow$ (i) : First proof.

Suppose that G is commutative. Then, Proposition 2.2 shows that G contains an element x whose order is the l.c.m. of the orders of the elements of G.

Suppose that G is not cyclic, that is, suppose order$(x) < n$. Let H be the subgroup of G generated by x and let y be an element of G not belonging to the subgroup H. If d is the order of y, then d divides order$(x) = \text{Card}\,(H)$ and H contains $\varphi(d)$ elements of order d. Thus G contains at least $\varphi(d)+1$ elements of order d, which contradicts condition (iii).

In this first proof, the groop G is supposed to be commutative. Here is a second proof, which works in the general case.

Second proof (Gauss).

Let d be a divisor of n, and let $\psi(d)$ be the number of elements of G whose order is equal to d. As shown by Lagrange, the order of every element of G divides its cardinality, thus

$$\text{(1)} \qquad \text{Card}\,(G) = n = \sum_{d|n} \psi(d)\,.$$

When $G = \mathbb{Z}/n\mathbb{Z}$, we have $\psi(d) = \varphi(d)$ for every divisor of n, then relation (1) becomes

$$n = \sum_{d|n} \varphi(d). \tag{2}$$

Thus, if property (iii) holds, that is, if $\psi(d) \leq \varphi(d)$ for every divisor d of n, then relations (1) and (2) imply

$$\psi(d) = \varphi(d), \qquad \text{for each divisor } d \text{ of } n.$$

In particular, we have $\psi(d) = \varphi(d)$. In other words, G contains $\varphi(d)$ elements of order n; thus G is cyclic! □

2. Application to G(p)

The preceding study leads to the following result.

THEOREM 2.1. — *If p is a prime number then the group $\mathrm{G}(p)$ is cyclic.*

Proof

Let p be a prime number, then $\mathrm{G}(p)$ contains $p-1$ elements. Let d be any divisor of $p-1$, then equation $X^d = 1$ has at most d solutions in the finite field $\mathbb{Z}/p\mathbb{Z}$. Thus condition (ii) of Proposition 2.3 is satisfied : $\mathrm{G}(p)$ is cyclic. □

An integer g of the interval $\{1, \ldots, p-1\}$ is called a *primitive element* modulo p if its class generates $\mathrm{G}(p)$. Here is the list of the smallest primitive elements modulo p for $p \leq 43$.

p	2	3	5	7	11	13	17	19	23	29	31	37	43
g	1	2	2	3	2	2	3	2	5	3	3	2	3

We verify it for $p = 17$: the order of each element is 16 or a strict divisor of 16 and

$$3^4 = 81 \equiv 13, \quad 3^8 \equiv 13^2 = 169 \equiv -1 \pmod{17},$$

thus the order of 3 modulo 17 is exactly 16.

Thus, if p is a prime number and if d divides $p-1$ then the group G(p) contains exactly $(p-1)/d$ elements of the form x^d, $x \in$ G(p). In particular, if p is odd then half of the elements of G(p) are squares; an integer a, nondivisible by p, whose class modulo p is a square, is called a *quadratic residue* modulo p, otherwise it is called a *non-residue* modulo the prime number p.

One defines the *symbol of Legendre*,

$$\left(\frac{a}{p}\right) = \begin{cases} +1 & \text{if } a \text{ is a quadratic residue mod } p, \\ -1 & \text{if } a \text{ is not a quadratic residue mod } p. \end{cases}$$

The symbol of Legendre satisfies the following properties.

THEOREM 2.2. — *Let p be an odd prime number and a and b be two integers nondivisible by p, then*

(i) $$\left(\frac{ab}{p}\right) = \left(\frac{a}{p}\right)\left(\frac{b}{p}\right),$$

(ii) $$\left(\frac{a}{p}\right) \equiv a^{(p-1)/2} \pmod{p}, \quad \text{(Euler)}.$$

Proof

Let G be a primitive element modulo p. Put $a \equiv g^h$ and $b \equiv g^k$. Then

$$\left(\frac{a}{p}\right) = (-1)^h, \quad \left(\frac{b}{p}\right) = (-1)^k \quad \text{and} \quad \left(\frac{ab}{p}\right) = (-1)^{h+k};$$

hence the first assertion.

Proof of (ii) : Each member of relation (ii) defines a morphism of G(p); this is obvious for the right-hand side, and, for the left one, it is the consequence of (i) [we consider that -1 belongs to G(p)]. Since G(p) is cyclic, it is sufficient to verify the relation when $a = g$. First,

$$\left(\frac{g}{p}\right) = -1.$$

Then, put $y = g^{(p-1)/2} \bmod p$, this number satisfies

$$y^2 \equiv g^{p-1} \equiv 1 \quad \text{and} \quad y \not\equiv 1 \pmod{p}.$$

This implies $y \equiv -1 \pmod{p}$. Hence the conclusion. □

If n is an odd integer and if a is an integer prime with n, the *symbol of Jacobi* is defined as follows :

$$\left(\frac{a}{n}\right) = \prod_{i=1}^{r} \left(\frac{a}{p_i}\right)^{\alpha_i} \quad \text{if} \quad n = {p_1}^{\alpha_1} \cdots {p_r}^{\alpha_r} \quad (p_i \text{ distinct}).$$

3. Structure of $\mathrm{G}(p^k)$, $k \geq 2$, *p odd prime*

We know that the group $\mathrm{G}(p^k)$ contains $\varphi(p^k) = (p-1)p^{k-1}$ elements and that the group $\mathrm{G}(p)$ is cyclic. Let a be an integer whose class modulo p generates $\mathrm{G}(p)$, then the order of a modulo p^k is a multiple of $p-1$. Moreover, consider the element $b = 1+p$, then

$$(*) \qquad b^{p^h} \equiv 1 + p^{h+1} \pmod{p^{h+2}}, \quad \text{for } h \in \mathbb{N}.$$

Indeed, this congruence is true for $h = 0$, and, if it is true at rank h, then

$$b^{p^h} = 1 + p^{h+1} + up^{h+2}, \quad u \in \mathbb{Z},$$

and

$$b^{p^{h+1}} = \left(1 + p^{h+1} + up^{h+2}\right)^p = 1 + p^{h+2} + vp^{h+3}, \quad v \in \mathbb{Z},$$

which shows that formula $(*)$ is still true at rank $h+1$. Hence $(*)$ holds for any h in $\mathbb{N}$. This formula, applied for $h = k-2$ and $h = k-1$ shows that the element b is of order p^{k-1} modulo p^k.

Now, it is possible to show that $\mathrm{G}(p^k)$ is cyclic. Indeed, there are two possible cases : either a generates the group $\mathrm{G}(p^k)$, or the element a does not generate $\mathrm{G}(p^k)$.

If the first case holds, there is nothing to prove. Suppose that a does not generate $\mathrm{G}(p^k)$. Then consider the element $x = a(1+p) \bmod p^k$, and let t be its order. Then $x \equiv a \pmod{p}$, and t is a multiple of $p-1$. Moreover, since a does not generate $G(p^k)$, then

$$a^{p^{k-2}(p-1)} \equiv 1 \pmod{p^k}.$$

Hence

$$x^{p^{k-2}(p-1)} \not\equiv 1 \pmod{p^k}.$$

Therefore, t is equal to $\varphi(p^k)$; in other words, x generates $\mathrm{G}(p^k)$.

We have proved the following theorem.

THEOREM 2.3. — *Let p be an odd prime number and let k be a rational integer ≥ 2. Then the group $G(p^k)$ is cyclic. Furthermore, if a is any integer whose class modulo p generates $G(p)$ then*

- *either a generates $G(p^k)$,*
- *or $a(1+p)$ generates $G(p^k)$.*

Examples

• $p = 3$, $k = 4$: Take $a = 2$, then $a^6 = 64 \equiv -17 \pmod{81}$, $a^{12} \equiv -35$ and finally $a^{18} \equiv 28$, which shows that the number 2 generates the group G(81).

• $p = 5$, $k = 3$: Take $a = 3$, we verify that $a^{20} \equiv 26 \pmod{125}$, which shows that 3 generates G(125).

4. Structure of G(2^k)

It has just been demonstrated that for any odd prime number p and any positive integer k, the group G(p^k) is cyclic. The case of $p = 2$ is special and the structure of the group G(2^k) depends on k.

If $k = 1$, then G$(2) = \{1\}$ is trivial. If $k = 2$, $G(4) = \{1, -1\}$ is cyclic, generated by -1. For $k \geq 3$, notice that for any odd integer x

$$x^{2^{k-2}} \equiv 1 \pmod{2^k}.$$

Indeed, this congruence is true for $k = 3$ (proof : for x odd, $x = 4u \pm 1$, hence $x^2 \equiv 1$ modulo 8) and if it holds at rank k for some odd integer x, then

$$x^{2^{k-2}} = 1 + v\,2^k \quad \text{and} \quad x^{2^{k-1}} = (1 + v\,2^k)^2 \equiv 1 \pmod{2^{k+1}},$$

which shows that for $k \geq 3$, the group G(2^k) is not cyclic.

The following lemma is somewhat similar to congruence $(*)$ of the previous paragraph.

LEMMA 2.1. — *For all integer $h \geq 0$, we get*

$$5^{2^h} \equiv 1 + 2^{h+2} \pmod{2^{h+3}}.$$

Proof

It is true for $h = 0$. Suppose it is demonstrated at rank h, then

$$5^{2^{h+1}} \equiv \left(1 + 2^{h+2}\,(1+2w)\right)^2 \pmod{2^{h+4}}.$$

Hence we have the conclusion. □

Applying the lemma for $h = k - 3$ shows that the class of 5 is of order 2^{k-2} in $\mathrm{G}\,(2^k)$. In other words, the number 5 generates a subgroup H of index 2 in group $\mathrm{G}\,(2^k)$.

Now, show that -1 does not belong to H. If it was true, then

$$5^h \equiv -1 \pmod{2^k}, \qquad \text{with} \quad 0 \le h < 2^{k-2},$$

hence $5^{2h} \equiv 1 \pmod{2^k}$, which implies $h = 2^{k-3}$, and the lemma gives a contradiction. Thus we have demonstrated the following result.

THEOREM 2.4. — *For $k < 3$ the group $\mathrm{G}\,(2^k)$ is cyclic. But for any integer $k \ge 3$ the group $\mathrm{G}\,(2^k)$ is not cyclic; it is composed of the classes of the elements $\pm 5^h$, $0 \le h < 2^{k-2}$, and hence it is isomorphic to the direct product*

$$(\mathbb{Z}/2^{k-2}\mathbb{Z}) \times \mathbb{Z}/2\mathbb{Z}.$$

5. Structure of $\mathrm{G}\,(n)$

Suppose that $n = p_1^{a_1} \cdots p_t^{a_t}$ is the decomposition of n, then

$$(*) \qquad \mathrm{G}\,(n) \sim \prod_{i=1}^{t} G(p_i^{a_i}).$$

If $\lambda(n)$ is the l.c.m. of the orders of elements of the group $\mathrm{G}\,(n)$, [this function was introduced by Carmichael†] the previous isomorphism implies

$$\lambda(n) = \text{l.c.m.}\left\{\lambda(p_1^{a_1}), \ldots, \lambda(p_t^{a_t})\right\}.$$

† R.D. Carmichael, *Bull. Amer. Math. Soc.*, 16, 1910, p. 232–238.

Moreover, the theorems above show that

$$\lambda(2^k) = 2^{k-2} \text{ if } \quad k \geq 3, \quad \lambda(2) = 1, \quad \lambda(4) = 2,$$

and

$$\lambda(p^k) = (p-1)\, p^{k-1},$$

if p is an odd prime number.

This leads to the computation of the function $\lambda(n)$. In particular, if n is an odd positive number, that factorizes as $n = p_1^{a_1} \cdots p_t^{a_t}$, then $\lambda(n)$ is given by the formula

$$\lambda(n) = p_1^{a_1 - 1} \cdots p_t^{a_t - 1} \text{ l.c.m.} (p_1 - 1, \ldots, p_t - 1).$$

Thus $\lambda(n)$ divides $2^{1-t}\varphi(n)$ for n odd.

Notice that the group $\mathrm{G}(n)$ is cyclic if, and only if, $\lambda(n) = \varphi(n)$. For n' odd and $n = n'$ or $n = 2n'$, this is equivalent to the fact that n' is a power of a prime number. For $n = 4n'$, for any integer n', we have $\lambda(n) < \varphi(n)$. Hence the following result :

Proposition 2.4. — *Let n be a* squarefree integer (*that is, nondivisible by the square of a prime number*). *Let a be any integer, and let e be an integer that satisfies*

$$e \equiv 1 \pmod{\lambda(n)},$$

then

$$a^e \equiv a \pmod{n}.$$

Proof

Thanks to the Chinese remainder theorem, we only have to consider the case of n prime. Then the proposition follows from $a^p \equiv a \pmod{p}$, which is a direct consequence of Fermat's theorem. □

2. Tests of primality

1. A general theorem

The following technical lemma will be very useful.

LEMMA 2.2. — *Let n be an odd integer, $n = p_1^{\alpha_1} \cdots p_r^{\alpha_r}$, and put $\nu = v_2(n-1)$, where $v_2(m)$ is the highest integer k, such that 2^k divides m. If $\nu' = \inf\{v_2(p_i - 1)\,;\, 1 \le i \le r\}$, then, always $\nu \ge \nu'$ and the equality holds if, and only if, the number of prime factors p_i of odd exponent and satisfying $v_2(p_i - 1) = \nu'$, is odd.*

Proof

Suppose the p_is ordered by increasing values of $v_2(p_i - 1)$. Then

$$n - 1 = \sum_{i=1}^{r} \left(p_i^{\alpha_i} - 1\right) p_{i+1}^{\alpha_{i+1}} \cdots p_r^{\alpha_r}.$$

Hence the inequality $\nu \ge \nu'$.

Moreover, suppose that the indices i for which $v_2(p_i - 1) = \nu'$ and α_i is odd are 1, 2, ..., k, then $2^{\nu'+1}$ divides the terms of the previous sum of indices $i > k$, and only those. Thus we have the relation

$$\nu - 1 = k\,2^{\nu'} \pmod{2^{\nu'+1}},$$

hence the second assertion. □

Consider the following properties of an integer n such that $n - 1 = 2^\nu n'$, where n' is odd :

$\mathrm{P}(n)$: n is prime,

$\mathrm{Q}(n)$: n is squarefree,

$\mathrm{F}(n) : \forall a, (a,n) = 1 \Longrightarrow a^{n-1} \equiv 1 \pmod{n}$,

$\mathrm{E}(n) : \forall a,\ (a,n) = 1 \implies a^{(n-1)/2} \equiv \pm 1 \pmod{n}$,

$S(n,a) : (a,n) = 1$ and $a^{(n-1)/2} \equiv \left(\frac{a}{n}\right) \pmod{n}$,

$\mathrm{S}(n) : \forall a,\ (a,n) = 1 \implies S(n,a)$,

$$\mathrm{M}(n,a) : (a,n) = 1 \quad \text{and}$$
$$a^{n'} \not\equiv 1 \pmod{n} \Longrightarrow \exists h,\, 0 \le h < n,\ a^{2^h} \equiv -1 \pmod{n}\,;$$

Moreover, define $\omega(n) = \sum_{p|n} 1$, that is the number of different prime numbers p that divide n.

THEOREM 5. — *Let n be an odd integer, $n-1 = 2^\nu n'$, n' odd. Then the following implications hold, (recall that p always designates a prime number)*

$$\mathrm{P}(n) \Longrightarrow \mathrm{E}(n) \Longrightarrow \mathrm{F}(n),$$
$$\forall a,\ (a,n) = 1 \Longrightarrow \Big(\mathrm{P}(n) \Longrightarrow \mathrm{M}(a,n)\Big),$$
$$\mathrm{M}(n,a) \Longrightarrow \mathrm{S}(n,a),$$
$$\mathrm{F}(n) \Longleftrightarrow \mathrm{Q}(n) \quad \textit{and} \quad \Big(p \mid n \Longrightarrow (p-1) \mid (n-1)\Big),$$
$$\mathrm{F}(n) \Longrightarrow \omega(n) \neq 2,$$
$$\mathrm{E}(n) \Longleftrightarrow \mathrm{P}(n) \quad \textit{or}$$
$$\Big(\ \mathrm{Q}(n) \text{ and } \{\ p \mid n \Longrightarrow \omega(p-1) \mid \tfrac{n-1}{2}\ \}\ \Big),$$

and, $$\mathrm{S}(n) \Longrightarrow \mathrm{P}(n).$$

Proof

$$\mathbf{P}\ (n) \Longrightarrow \mathbf{E}\ (n) \Longrightarrow \mathbf{F}\ (n) : \text{Obvious.}$$

$$\forall a,\ (a,n) = 1 \Longrightarrow \big(\mathbf{P}(n) \Longrightarrow \mathbf{M}(n,\, a)\big)$$

Indeed, let a be an integer that is relatively prime with n, where n is prime, and such that we have the relation $a^{n'} \not\equiv 1 \pmod{n}$; then there is an integer h, with $0 \le h < n$, such that

$$y := a^{2^h n'} \not\equiv 1 \quad \text{and} \quad a^{2^{h+1} n'} \equiv 1 \pmod{n}.$$

Then, modulo n, the integer y satisfies $y^2 \equiv 1$ and $y \not\equiv 1$, and, since n is prime, $y \equiv -1 \pmod{n}$.

$\mathbf{M}\ (n,a) \implies \mathbf{S}\ (n,a)$

Let a be an integer relatively prime with n. Put

$$\chi(a) = a^{(n-1)/2}\left(\frac{a}{n}\right) \mod n.$$

Show that $\mathrm{M}(n,a)$ implies $\chi(a)=1$.

1°) Suppose $a^{n'} \equiv 1 \pmod{n}$, then

$$a^{(n-1)/2} \equiv 1 \pmod{n}$$

and for each prime divisor p of n

$$\left(\frac{a}{p}\right) = \left(\frac{a}{p}\right)^{n'} = 1;$$

thus we have $\chi(a)=1$.

2°) Suppose now that $a^{n'} \not\equiv 1 \pmod{n}$ and that $\mathrm{M}(n,a)$ is true. Then,

$$a^{2^b n'} \equiv -1 \quad \text{and} \quad a^{2^{b+1} n'} \equiv 1, \qquad \text{with } 0 \le h < v.$$

As above, let $n = {p_1}^{\alpha_1} \cdots {p_r}^{\alpha_r}$ be the decomposition of n into prime factors. According to Euler's formula,

$$h = v_2(p_i - 1) - 1 \implies \left(\frac{a}{p_i}\right) = -1, \text{ and } h < v_2(p_i - 1) - 1 \implies \left(\frac{a}{p_i}\right) = 1.$$

First, suppose that $h = \nu - 1$. This corresponds to the case $\nu = \nu'$ of the previous lemma, which shows that we get (with the notations of the demonstration of this lemma),

$$\left(\frac{a}{p_i}\right) = 1 \iff i = k+1, \dots, r.$$

Moreover, the sum $\alpha_1 + \cdots + \alpha_k$ is odd, thus

$$\left(\frac{a}{n}\right) = \left(\frac{a}{p_1}\right)^{\alpha_1} \cdots \left(\frac{a}{p_k}\right)^{\alpha_k} \left(\frac{a}{p_{k+1}}\right)^{\alpha_{k+1}} \cdots \left(\frac{a}{p_r}\right)^{\alpha_r} = -1.$$

By the definition of h, we have $a^{(n-1)/2} = -1$, thus $\chi(a) = 1$.

There remains the case where $h < \nu - 1$. Then

$$a^{(n-1)/2} = 1 \quad \text{and} \quad \left(\frac{a}{p_i}\right) = 1 \quad \text{for} \quad i = 1, 2, \ldots, r.$$

Thus $\chi(a) = 1$ in all cases.

$$\mathbf{F}\ (n) \iff \mathbf{Q}\ (n) \quad \text{and} \quad \Big(p|n \implies (p-1)\,|\,(n-1)\Big)$$

If n is not squarefree, let p be a prime number such that $n = p^\alpha m$, where $\alpha \geq 2$ and p does not divide m. Then $\mathrm{G}(n) \sim \mathrm{G}(p^\alpha) \times \mathrm{G}(m)$ and the group $\mathrm{G}(n)$ contains an element a of order p. Since p does not divide $n-1$, the congruence

$$a^{n-1} \equiv 1 \pmod{n}$$

is impossible. Hence the implication $\mathrm{F}(n) \implies \mathrm{Q}(n)$ is proved.

If p divides n, then $\mathrm{G}(n)$ contains an element of order $p-1$, and, consequently, if property $\mathrm{F}(n)$ holds, then $p-1$ divides $n-1$. The inverse implication is obvious.

$$\mathbf{F}\ (n) \implies \omega(n) \neq 2$$

Now suppose $n = pq$, where p and q are odd prime numbers, $p > q$. Then $(p-1)$ divides $(n-1)$. Also

$$(p-1) \mid (p-1)(q-1).$$

Thus, by difference,

$$p + q - 2 = s(p-1),$$

with $s \geq 2$. Hence $q \geq p$, contradiction.

$$\mathbf{E}\ (n) \iff \mathbf{P}(n) \ \text{ or } \ \Big(\ \mathrm{Q}(n) \text{ and } \{p|n \implies (p-1) \mid (\tfrac{n-1}{2})\}\ \Big)$$

Suppose that $\mathrm{E}(n)$ is true and that n is not prime. Then n must be squarefree.

Let p be a prime divisor of n, since $\mathrm{G}(p)$ is cyclic and since p and n/p are relatively prime, there exists an element a of $\mathrm{G}(n)$ such that

$$\text{order } (a \bmod p) = p - 1 \quad \text{and} \quad a \equiv 1 \pmod{n/p}.$$

Then

$$\left\{a^{(n-1)/2} \equiv \pm 1 \pmod n \text{ and } a \equiv 1 \pmod{n/p}\right\} \Longrightarrow a^{(n-1)/2} \equiv 1,$$

hence $p-1$ divides $(n-1)/2$.

Again, the inverse implication is obvious.

$$\mathbf{S}\ (n) \Longrightarrow \mathbf{P}\ (n)$$

Suppose that n is nonprime (say $n = p_1 \cdots p_r$, $r \geq 2$) and that $\chi(a) = 1$ for any element a of $\mathrm{G}(n)$. We know that the prime numbers p_i are two-by-two distinct. Let a be an element of $\mathrm{G}(n)$ such that

$$\left(\frac{a}{p_1}\right) = -1 \quad \text{and} \quad a \equiv 1 \pmod{p_2 \cdots p_r}.$$

Then the hypothesis $\chi(a) = 1$ implies

$$a^{(n-1)/2} \equiv -1 \pmod n,$$

which contradicts the condition $a \equiv 1 \pmod{p_2}$. This is the end of the demonstration. □

2. A simple test of primality

The previous theorem gives a very simple test of primality :

THEOREM 2.6. — *Let n be an odd integer, $n > 1$. Then the number n is prime if, and only if, the following two conditions hold simultaneously*

$$\left\{\forall a, (a,n) = 1 \Longrightarrow a^{(n-1)/2} \equiv \pm 1 \bmod n\right\}$$

and

$$\left\{\exists a, a^{(n-1)/2} \equiv -1 \bmod n\right\}.$$

Proof

Call $C(n)$ the conjonction of the two conditions considered in the statement of Theorem 2.6. It is obvious that condition $C(n)$ is necessary. Show that it is sufficient. For that we argue *ad reductio ad absurdum* : we suppose that some odd integer $n \geq 3$ satisfies $\mathrm{C}(n)$ without being prime. Theorem 2.5 shows that n is squarefree, since

$$\mathrm{C}(n) \Longrightarrow \mathrm{E}(n) \Longrightarrow \mathrm{F}(n) \Longrightarrow \mathrm{Q}(n).$$

Suppose that x is some integer such that $x^{(n-1)/2} \equiv -1 \pmod{n}$ and that p is a prime divisor of n, $n = pn'$ with $n' > 1$ [and $(p, n') = 1$]. According to the Chinese remainder theorem, there is a rational integer a that satisfies the two conditions $a \equiv x \pmod{p}$ and $a \equiv 1 \pmod{n'}$.

Then

$$a^{(n-1)/2} \equiv x^{(n-1)/2} \equiv -1 \pmod{p} \quad \text{and} \quad a^{(n-1)/2} \equiv 1 \pmod{n'},$$

which contradicts the condition $a^{(n-1)/2} \equiv \pm 1 \pmod{n}$. And the result follows. ▯

The test above is not very useful in practice : if there exists a rational integer a such that $a^{(n-1)/2} \not\equiv \pm 1$ modulo n, then we know that n is not prime; but while $a^{(n-1)/2} \equiv 1 \pmod{n}$, we cannot reach a conclusion.

Notice also that there are nonprime odd numbers n such that any rational integer a coprime with n satisfies $a^{(n-1)/2} \equiv 1 \pmod{n}$. The two smallest such numbers are $1729 = 7 \times 13 \times 19$ and $2465 = 5 \times 17 \times 29$; it is not known yet whether there is an infinity of such numbers. Obviously, such numbers are Carmichael numbers (defined in Chaptre 1, Section 11), but the converse is false (consider the number 561). Again, it is not whether the class of those numbers is finite or not.

3. Elementary tests

Let n be an odd integer ≥ 3. We want to know whether n is or is not prime. We now deduce several tests of primality from Theorem 2.5.

1. Test of Solovay-Strassen†

Verify that $\chi(a) = 1$ for some values of a, for example for the choices $a = 2, 3, 5, 7, 11\ldots$ If there is a value such that $\chi(a) \neq 1$, then it is proved that n is not prime. Otherwise, there must be a stop test, which will be treated later. Notice that if n is not prime then the character χ is not trivial and its kernel is of index in $\mathrm{G}(n)$ at least equal to two, which shows that at most half of the elements a of the group $\mathrm{G}(n)$ satisfy $\chi(a) = 1$.

---

† R. Solovay and V. Strassen, A fast Monte-Carlo primality test, *SIAM J. Computing*, 6, 1977, p. 84–85; 7, 1978, p. 118.

2. Test of Miller¶-Rabin‡

Suppose $n-1=2^\nu n'$, with n' odd. For some positive integers a, the expression $b=a^{n'} \bmod n$ is evaluated, and if $b\neq 1$, compute successively b^2, b^4, b^8, …, b^h, … where the exponent h satisfies $h\leq 2^{\nu-1}$, until a value congruent to -1 modulo n has been found. The situation is the same than previously : if no value $\equiv -1$ has been found; then n is not prime, otherwise, try another value a.

Theorem 2.5 points out that property $\mathrm{M}(n,a)$ implies $S(n,a)$, thus this test is at least as strong as the test of Solovay-Strassen.

3. Study of the cost of these tests

Let N be the number of tries done by either one of the two previous tests, then the number of elementary operations is in $\mathrm{O}(N\ \mathrm{Log}\, n)$ and the cost (using algorithms of ordinary multiplication) is in $\mathrm{O}\big(N\,(\mathrm{Log}\, n)^3\big)$.

Suppose n is not prime and that one of these tests is done for the successive prime numbers $a=2$, 3, 5, 7, 11, …, and let A be the smallest positive integer that satisfies both $(A,n)=1$ and $\chi(A)\neq 1$, then

$$N\leq A \quad \big[\text{ and even } N=\mathrm{O}(A/\mathrm{Log}\, A)\ \big].$$

We still have to bound A, but it is difficult. The results in this field are :

- thanks to the works of Weil, Vinogradov, and Burgess||, footnote we know that A satisfies

$$A=\mathrm{O}\left(n^{1/4\sqrt{e}}\right);$$

- Ankeny and Montgomery have shown that, if the hypothesis of Riemann is true then

$$A\leq c\,(\mathrm{Log}\, n)^2.$$

¶ G.L. Miller, Riemann's hypothesis and tests for primality, *J. Comp. Syst. Sci.*, 13, 1976, p. 300–317.

‡ M.O. Rabin, Probability Algorithms, in *Algorithms and Complexity*, ed. by J.F. Traub, Academic Press, New York, 1976, p. 35–36.

|| D.A. Burgess .— On character sums and L–series, *Proc. London Math. Soc.*, 12 (3), 1962, p. 193–206.

Explicit values of this constant c have been computed by Oesterlé and Weinberger, the best published value , $c = 2$, has been obtained by Bach§

According to the works of Adleman, Pomerance and Rumely*, and those of Cohen and Lenstra**, the primality of an integer n can be tested in a time at most

$$\mathrm{O}\left((\operatorname{Log} n)^{c \operatorname{Log}\operatorname{Log}\operatorname{Log} n}\right),$$

where c is some positive constant. But the method is too complex to be detailed here.

4. Statistics on $\mathrm{G}(n)$

The following result specifies Theorem 2.5.

PROPOSITION 2.5. — *Let n be an odd number that is neither a square nor a power of a prime number, say*

$$n = p_1^{e_1} \cdots p_r^{e_r}, \quad r \geq 2.$$

Write

$$F = \{x \in \mathrm{G}(n)\,;\ x^{n-1} \equiv 1 \pmod n\},$$
$$E = \{x \in \mathrm{G}(n)\,;\ x^{(n-1)/2} \equiv \pm 1 \pmod n\},$$
$$H = \{x \in \mathrm{G}(n)\,;\ \chi(x) = 1\}, \quad (\textit{previous notations}).$$

Then F, E, and H are subgroups of $\mathrm{G}(n)$ *that satisfy*

$$F \supset E \supset H,$$
$$E = F \implies [E:H] = 2,$$
$$\operatorname{Card}(F) = d_1 \cdots d_r \quad \textit{where} \quad d_i = \text{g.c.d.}\,(n-1, p_i - 1).$$

---

§ E. Bach .— Analytic methods in the Analysis and Design of Number-Theoretic Algorithms, *A.C.M. Distinguished Dissertations*, MIT Press, Cambridge (Mass.), 1985.

* L. Adleman, C. Pomerance and R. Rumely .— On distinguishing prime numbers from composite numbers, *Ann. of Math., 117, 1983, p. 173–206.*

** H. Cohen and H.W. Lenstra .— Primality testings and Jacobi sums, *Math. Comp.*, 42, 1984, p. 297–330.

Proof

Put $G = \mathrm{G}(n)$ to simplify the notations. The congruence $x^{n-1} \equiv 1$ modulo n is equivalent to the conditions

$$x^{n-1} \equiv 1 \pmod{p_i^{e_i}}, \qquad \text{for } i = 1, \ldots, r.$$

Since each group $\mathrm{G}(p_i^{e_i})$ is cyclic, it contains

$$d_i = \text{g.c.d.}\left(n-1, {p_i}^{e_i-1}(p_i-1)\right) = \text{g.c.d.}\,(n-1, p_i-1)$$

solutions of the congruence

$$x^{n-1} \equiv 1 \pmod{p_i^{e_i}},$$

hence the formula giving the cardinality of F.

Let t be the number of indices i such that there exists $x \in G$ satisfying

$$x^{(n-1)/2} \equiv -1 \pmod{p_i}.$$

There are the three following possibilities

(i) $t = 0$ and then

$$E = F = \{x \in G\,;\, x^{(n-1)/2} \equiv 1 \pmod{n}\},$$

(ii) $0 < t < r$, then

$$E = \{x \in G\,;\, x^{(n-1)/2} \equiv \pm 1 \pmod{n}\} \quad \text{and} \quad [F : E] = 2^t,$$

(iii) $t = r$ and then the obvious relation

$$E = \{x \in G\,;\, x^{(n-1)/2} \equiv 1 \pmod{n}\} \cup \{x \in G\,;\, x^{(n-1)/2} \equiv -1 \pmod{n}\}$$

shows that $[F : E] = 2^{r-1}$.

For the group H the possibilities are the two following cases :

(a) if $E = \{x \in G\,;\, x^{(n-1)/2} \equiv 1 \pmod{n}\}$, cases (i) and (ii), then

$$\chi(n) = \left(\frac{x}{n}\right)$$

and, since n is not a square, the symbol of Jacobi is not trivial and

$$[E : H] = 2, \qquad \text{thus} \qquad [F : H] \geq 2,$$

(b) otherwise, case (iii) holds and $[F:H] \geq 2^{r-1} \geq 2$.

Examine the consequences of the previous remarks. Let n be a nonprime integer.

• If n is not squarefree, then $[G:F] \geq 3$, and thus $[G:H] \geq 6$, (Prove this inequality as an exercise).

• If n is squarefree, $[G:H] \geq 4$ unless $G = E$, that is, unless n satisfies the condition

$$\forall a \left\{ (a,n) = 1 \implies a^{(n-1)/2} \equiv 1 \pmod{n} \right\}.$$

In this last case, $n \equiv 1 \pmod 4$. Then put

$$H' = \left\{ x \in G \, ; \, x^{(n-1)/2^{\mu}} \equiv \pm 1 \pmod{n} \right\}, \quad \text{where} \quad \mu = \nu - \nu' - 1,$$

and then each integer x that satisfies Miller-Rabin's test belongs to H'. The same argument as above shows that

$$[G:H'] = 2^{r-1} \geq 4.$$

Consequently, in any case

$$\text{Card} \left\{ x \in G \, ; \, x \text{ verify} \, \mathrm{M}(x,n) \right\} \leq \tfrac{1}{4} \, \text{Card} \left\{ \mathrm{G}(n) \right\}.$$

3. Factorization of rational integers

In this section, we discuss how to factorize a positive integer n (that is, to decompose this integer into a product of prime factors, cf. Theorem 1.10). Without any loss of generality, we suppose n odd and $n \geq 3$.

1. Methods by successive divisions

The usual scholar method is a particular case of the following method : successively perform the euclidean division of n by the elements $d_1, \ldots, d_k$ of an increasing sequence of rational integers that contains the sequence of prime numbers.

For example take the following choices of the sequences (d_k) :

- the sequence of odd numbers 3, 5, 7, 9, 11, 13, 15, ..., which leads to a very simple algorithm, but which obviously holds a number of useless divisions;
- the sequence of the odd prime numbers 3, 5, 7, 11, 13, 17, ..., which is optimal for the number of divisions, but which needs to have — or to construct — a list of all the prime numbers at most equal to $\sqrt{n}$;
- for example the sequences 3, 5, 7, 11, 13, 17, 19, 23, 29, 31, 37, 41, 43, 49, ... of the numbers 3 and 5 followed by the odd numbers neither divisible by 3 nor by 5.

The third possibility has the following advantages :

1°) It avoids many of the useless operations of the first method; for example, by eliminating the strict multiples of 3 in a list, we save 33% of the divisions of the first version of the algorithm, by eliminating the odd numbers nonprime with 15 (except 3 and 5) we save more than 46% of the operations.

More generally, by eliminating the numbers nonprime with a given integer k, we save a number of operations equal to $\big(k - \varphi(k)\big)/k$. For $k = 105$, this proportion is above 54%.

2°) It is not necessary to stock the sequences (d_n) : this sequence is equal to a combination of arithmetic progressions and can easily be built by "seaving".

As an example, take the algorithm for the search of the smallest prime factor of n. If $\mathrm{P}(n)$ now designates the biggest prime factor of n (we do not keep the notations used for Theorem 2.5), then the number of operations of this algorithm is in $\mathrm{O}\big(\mathrm{P}(n)\big)$. In the worst case — that is when the integer n is the product of two distinct prime numbers close to $\sqrt{n}$ — this number of operations is approximately $\sqrt{n}$.

2. Method of Fermat§

The idea behind this method is to replace a factorization $n = ab$, where $a \geq b$, by the decomposition $n = x^2 - y^2 = (x + y)(x - y)$, where

§ *See L.E. Dickson .— History of the Theory of Numbers, Chelsea, New York, 1952, p. 357.*

```
data : n integer ≥ 2 ;
m := n ; i := 0 ;
while m mod 2 = 0 do
  begin
  |   i := i + 1 ; T[i] := 2 ; m := m div 2
  end ;
a := 3 ;
while a² ≤ m do
  begin
  |   while m mod a = 0 do
  |     begin
  |     |   i := i + 1 ; T[i] := a ; m := mdiv a
  |     end ;
  |   a := a + 2
  end ;
if m > 1 then
  begin
  |   i := i + 1 ; T[i] := m
  end.
```

Factorization by divisions

$x = (a+b)/2$ and $y = (a-b)/2$. The procedure is the following. For $x = [\sqrt{n}], [\sqrt{n+1}], \ldots$, compute $x^2 - n$ until a square is found.

Suppose n is equal to $n = pq$, with $p \leq q$ and $p \leq n^\alpha$ (thus $\alpha \leq 1/2$). Then the computation must continue until $x > n^{1-\alpha}/2$. Thus for α fixed, with $\alpha < 1/2$, the cost of this method is not in $\mathrm{O}(\sqrt{n})$.

Therefore, the method of Fermat is only useful if the integer n contains a factor close to $\sqrt{n}$.

3. Method of Sherman Lehman¶

The idea behind this method belongs to Legendre : if there exist two rational integers a and b such that $a^2 \equiv b^2 \pmod{n}$ and $a \not\equiv \pm b \pmod{n}$, then the g.c.d. of $a+b$ and n is a nontrivial factor of n.

¶ *R. Sherman Lehman .— Factoring large integers, Math. Comp., 28,* 1974, *p. 637–646.*

The procedure is the following :

(i) Search for the prime divisors of n until $n^{1/3}$.

If no factor has been found then the integer n is equal to $n = pq$ with

$$n^{1/3} < p \leq q < n^{2/3}$$

(we suppose that n is nonprime).

(ii) For $k = 1, 2, \ldots, [n^{1/3}]$ and $d = 0, 1, \ldots, 1 + [n^{1/6}/(4\sqrt{k})]$ test whether the number

$$\left\{ \left[(4\,k\,n)^{1/2} \right] + d \right\}^2 - 4\,k\,n$$

is a square. If so, we have found integers a and b as above. Since $a + b$ and $a - b$ are both in the range $\left]1, n\right[$ (for $n > 6$), it gives a factorization of n.

Now, show that, if $n = pq$ with $n^{1/3} < p \leq q < n^{2/3}$, then step (ii) always gives the factors p and q.

We use the following property

$$(*) \qquad \exists r, s \geq 1, rs < n^{1/3} \quad \text{and} \quad |ps - qr| < n^{1/3},$$

which will be demonstrated later.

Then, if $k = rs$, we have the relation

$$4kn = (ps + qr)^2 - (ps - qr)^2,$$

which shows that $(ps + qr)^2 - 4kn$ is a square $< n^{2/3}$.

Let $d = ps + qr - [2\sqrt{kn}]$, then

$$n^{2/3} > (ps + qr)^2 - 4kn > 4\,(d-1)\,\sqrt{kn},$$

thus $d < n^{2/3}(4\sqrt{kn})^{-1} + 1$. And we get the factorization at step (ii).

Now, we demonstrate property $(*)$.

Let $(r_i/s_i)_{i \geq 1}$ be the sequence of the principal convergents of the rational number q/p. Then the first fraction of this sequence is defined by the formulas $r_1 = [q/p]$ and $s_1 = 1$; which implies $0 < r_1 s_1 < n^{1/3}$.

Let m be the biggest integer, such that $r_m s_m < n^{1/3}$. Then put $r = r_m$ and $s = s_m$ to simplify the notations. We get $r \geq 1$, $s \geq 1$ and $rs < n^{1/3}$, so that

$$|rp - sq| = ps\left|\frac{r}{s} - \frac{q}{p}\right| \leq ps\,\frac{1}{ss_{m+1}} = \frac{p}{s_{m+1}}.$$

If $p/s_{m+1} \leq q/r_{m+1}$ then

$$|rp - qs| \leq \left(\frac{pq}{r_{m+1}s_{m+1}}\right)^{1/2} < n^{1/3}.$$

If $p/s_{m+1} > q/r_{m+1}$, then s/r and s_{m+1}/r_{m+1} are principal convergents of the rational number p/q and the result can be demonstrated in the same way.

It is obvious that the number of operations of this method is in $\mathrm{O}(n^{1/3})$.

4. Pollard's rho method†

In Pollard's method, we compute a sequence (x_n) defined by the formulas

$$x_1 = c \quad \text{and} \quad x_h = ({x_{h-1}}^2 + 1) \bmod n \quad \text{for } h \geq 2.$$

Let p be the smallest prime divisor of n. Then the sequence of the numbers $x_h \bmod p$ is ultimately periodical and thus (exercise) there exists a positive integer k such that

$$x_{2k} \equiv x_k \pmod{p},$$

and then p divides the g.c.d. of the integers $(x_{2k} - x_k)$ and n.

If this g.c.d. is not equal to n, we get a nontrivial divisor of n. Moreover, if the sequence $x_h \bmod p$ behaves like a random sequence, then

$$x_{2k} \equiv x_k \pmod{p}$$

for some k bounded above by $\mathrm{O}\left(\sqrt{p}\right)$.

† *J.M. Pollard .— A Monte-Carlo Method for Factorization Algorithm, Nordisk Tidskrift för Informationsbehandling (BIT), 15, 1975, p. 331–334.*

Exercise

Demonstrate this result. [Compute the probability for all the elements $x_1, x_2, \ldots, x_i$ to be distinct].

Experimentally, this method almost always gives a nontrivial factorization of n in a number of operations in $\mathrm{O}(p^{1/2})$, where p is the smallest prime divisor of n. Thus — in general — this method gives a nontrivial factor of n, for n nonprime, in a time $\mathrm{O}(n^{1/4})$.

Remark

The main point of this method is the choice of the sequence (x_n) mod p, which, in some sense, should behave like a random sequence. If we take a sequence satisfying a linear recurrence, this method fails (this can be proved). Nevertheless, for the choice of a sequence of integers (x_h) which satisfies $x_h \equiv ({x_{h-1}}^2+1) \bmod p$ — or, more generally, which satisfies $x_h \equiv ({x_{h-1}}^2+a) \bmod p$ — the practice shows that the method works, but this has not yet been strictly demonstrated.

Exercises

1. Let a be an integer, $a \geq 2$. Prove that there exists an infinity of positive integers n such that $a^{n-1} \equiv 1 \pmod{n}$.

[Let p be a prime number that does not divide the product $a(a^2-1)$, and take $n = (a^{2p}-1)/(a^2-1)$. Demonstrate that the product $2p$ divides $n-1$ and that a satisfies $a^{2p} \equiv 1 \bmod n \ldots$]

2. Let p be a prime odd number and h an integer that satisfies the inequality $2^h < p$. Suppose that n is equal either to $hp+1$ or to hp^2+1 and that

$$2^h \not\equiv 1 \pmod{n} \quad \text{but} \quad 2^{n-1} \equiv 1 \pmod{n}.$$

Prove that n is prime.

[First, demonstrate that, if n is not prime, then it has a prime factor q with $q \equiv 1 \pmod{p}$.]

3. Let n be an integer such that $n-1 = p^a r$ with $a \geq 1$ and r nondivisible by p, where (as usual) p is a prime number.

1°) Suppose that there is an integer x such that

$$\left(x^{(n-1)/p} - 1, n\right) = 1 \quad \text{and} \quad x^{n-1} \equiv 1 \pmod{n}.$$

Prove that if q is a prime divisor of n then p^a divides $q-1$.

2°) Suppose that n is equal to

$$n = {p_1}^{a_1} \cdots {p_r}^{a_r} Q \quad \text{with} \quad {p_1}^{a_1} \cdots {p_r}^{a_r} > Q$$

and that, for each index $i = 1, \ldots, r$, there is an integer x_i that satisfies

$$\left({x_i}^{(n-1)/p_i} - 1, n\right) = 1 \quad \text{and} \quad {x_i}^{n-1} \equiv 1 \pmod{n}.$$

Show that n is prime.

4. Let n be an odd positive integer. Demonstrate that the following four properties are equivalent :

(i) n is prime,

(ii) $\mathrm{G}(n)$ contains an element of order $> (n-1)/2$,

(iii) $x^2 \equiv 1 \pmod{n} \implies x \equiv \pm 1 \pmod{n}$,

(iv) if $n - 1 = 2^r n'$, n' odd, then

$$\left(a \in \mathrm{G}(n) \text{ and } a^{n'} \neq 1\right) \implies \exists s,\ 0 \le s < r,\ a^{2^s n'} = -1.$$

Which of these properties can be useful to test primality?

5. For a nonnegative integer n, the *number of Fermat*, F_n, is defined by the formula

$$F_n = 2^{2^n} + 1.$$

1°) Demonstrate that if a number $a^n + 1$ is prime, such that $a \ge 2$, then a is even and n is equal to a power of 2.

2°) Demonstrate the following properties :

(i) for all $r > 0$, F_n divides $F_{n+r} - 2$,

(ii) for $0 \le m < n$, the integers F_m and F_n are relatively prime,

(iii) F_n is prime for $n \le 4$,

(iv) 641 divides F_5.

6. We have seen that for $k \ge 3$ any element a of $\mathrm{G}(2^k)$ satisfies one and only one congruence of the form

$$a \equiv (-1)^e\, 5^f \pmod{2^k},\ e \in \{0,1\}, \quad \text{and} \quad f \in \{0, 1, \ldots, 2^{k-2} - 1\}.$$

Deduce that an odd integer a is a square modulo 2^k, for $k \ge 3$, if, and only if, this integer satisfies $a \equiv 1 \pmod{8}$.

7. Criterion of Gauss

Let p be an odd prime number and let a be any integer not divisible by the prime p. Write

$$\nu = \mathrm{Card}\Big\{k\,;\ 1 \le k \le \tfrac{p-1}{2} \text{ and } \exists s,\ 1 \le s \le \tfrac{p-1}{2},\ ka \equiv -s \pmod p\Big\}.$$

Demonstrate the relation

$$\left(\frac{a}{p}\right) = (-1)^{\nu}.$$

Deduce that

$$\left(\frac{2}{p}\right) = +1 \iff p \equiv \pm 1 \pmod 8.$$

8. Symbol of Jacobi

Let b and b' be two odd integers.

1°) Demonstrate that the symbol of Jacobi satisfies the following properties.

(i) $$\left(\frac{aa'}{b}\right) = \left(\frac{a}{b}\right)\left(\frac{a'}{b}\right),$$

(ii) $$\left(\frac{a}{bb'}\right) = \left(\frac{a}{b}\right)\left(\frac{a}{b'}\right),$$

(iii) $$\left(\frac{-1}{b}\right) = (-1)^{\frac{b-1}{2}}, \qquad \left(\frac{2}{b}\right) = (-1)^{\frac{b^2-1}{8}},$$

[For the second formula, use the previous exercise.]

(iv) $$\left(\frac{a}{b}\right) = (-1)^{\frac{a-1}{2}\,\frac{b-1}{2}}\left(\frac{b}{a}\right),$$

for a odd, $b \ge 3$. [Admit the law of quadratic reciprocity

$$\left(\frac{p}{q}\right) = (-1)^{\frac{p-1}{2}\,\frac{q-1}{2}}\left(\frac{q}{p}\right),$$

when p and q are two distinct prime odd numbers. A proof of the law of quadratic reciprocity is given in Chapter 6, Exercise 21.]

2°) Deduce an algorithm to compute the symbol of Jacobi from the algorithm of Euclid. Compute the cost of this algorithm and use it to compute the symbol $\left(\frac{753}{811}\right)$.

9. Let F_n be the nth number of Fermat (see Exercise 5 above). Demonstrate that F_n divides the integer $2^{F_n} - 2$. Verify that the number 341 is not prime but still divides $2^{341} - 2$.

10. Let p be an odd prime number. Thanks to the criterion of Gauss (see Exercise 7), demonstrate the following properties :

(i) $\left(\frac{3}{p}\right) = +1$ if, and only if, $p \equiv \pm 1 \pmod 6$ (for $p \neq 3$),

(ii) $\left(\frac{5}{p}\right) = +1$ if, and only if, $p \equiv \pm 1 \pmod 5$ (for $p \neq 5$).

11. Let x and y be two integers. Demonstrate the following implications :

$$x^2 - y^2 \equiv 1 \pmod 3 \implies 3 \mid y,$$

$$x^2 - y^2 \equiv 1 \pmod 5 \implies 5 \mid xy.$$

12. If p is an odd prime number, the number $M = M_p = 2^p - 1$ is called a *number of Mersenne*. This exercise gives a criterion of primality of these numbers, the *criterion of Lucas*.

1°) Demonstrate that if the number $a^n - 1$ is prime then $a = 2$ and n is prime.

2°) Determine the integers $p \leq 17$ such that M_p is prime. [Answer : $p \neq 11$.]

3°) Let ω be the *golden number*, $\omega = (1 + \sqrt{5})/2$, and let p be a prime odd number. With the help of Exercise 10(ii), demonstrate that

(i) $$p \equiv \pm 1 \pmod 5 \implies \omega^{p-1} \equiv 1 \pmod p,$$

and

(ii) $$p \equiv \pm 2 \pmod 5 \implies \omega^{2(p+1)} \equiv 1 \pmod p.$$

[In each of these two cases, consider the number $\omega^p \bmod p$.]

4°) Let $\omega' = (1-\sqrt{5})/2$ be the conjugate of the golden number. For n positive, write

$$r_n = \omega^{2^n} + \omega'^{2^n}.$$

Then demonstrate that r_n is an integer, and demonstrate that, for m and n distinct, the integers r_m and r_n are relatively prime.

5°) Demonstrate the following implication :

$$M_p \text{ prime} \implies r_{p-1} \equiv 0 \pmod{M_p}.$$

6°) We now want to demonstrate the converse of the previous implication. Consider the property

$$r_{p-1} \equiv 0 \pmod{M_p} \implies M_p \text{ prime}.$$

(i) If p' is a prime divisor of M_p, with $p' \equiv \pm 1 \pmod 5$, demonstrate that

$$2^{p+1} \mid (p'-1).$$

[Use question 3 °(i).]

Deduce that M_p does not admit a divisor of this type.

(ii) If q' is a prime divisor of M_p, with $q' \equiv \pm 2 \pmod 5$, demonstrate that

$$2^p \mid (q'+1).$$

Deduce that M_p admits at most one divisor of this type.

(iii) Conclude.

7°) Having demonstrated the criterion of Lucas :

$$M_p \text{ prime} \iff r_{p-1} \equiv 0 \pmod{M_p},$$

obtain the cost of this test.

13. Let a and b be two coprime integers. Consider the sequence of integers (u_n) defined by

$$u_0 = 1, \quad u_1 = 1, \quad u_{n+2} = au_{n+1} - bu_n \quad \text{for } n \geq 0 .$$

Let m be a prime odd integer that is coprime with $a^2 - 4b$. Prove that if m divides u_{m+1} but does not divide any of the numbers $u_{(m+1)/p}$, for p prime divisor of $m+1$, then the number m is prime.

14. This exercise demonstrates the following property : for a rational a, with $a \neq 1$, there exists an infinity of integers n such that the number $S_n = 2^{2^n} + a$ is not prime. We follow a demonstration given by A. Schinzel.

Consider a fixed number S_n, where n is arbitrarily big and suppose that this number S_n is prime, say $S_n = p$. Write $p - 1 = 2^s\,h$, with h odd, and put $\varphi(h) = k$.

1°) Demonstrate the congruences

$$2^k \equiv 1 \pmod{h} \quad \text{and} \quad 2^{t+k} \equiv 2^t \pmod{p-1}, \text{ for all } t \geq s .$$

2°) Using Fermat's theorem, demonstrate that S_{n+k} is divisible by p. Conclude that the number S_{n+k} is composite.

15. Estimates on prime numbers : lower bounds

For a positive rational integer $n \geq 2$, we define $d_n = \text{l.c.m.}\{1, 2, \dots, n\}$.

1°) For $1 \leq m \leq n$, put

$$I(m,n) = \int_0^1 x^{m-1}\,(1-x)^{n-m}\,dx .$$

(i) Demonstrate the formula

$$I(m,n) = \sum_{j=0}^{m-n} (-1)^j \binom{n-m}{j} \frac{1}{m+j} \cdot$$

(ii) Prove that $d_n\, I(m,n)$ is a rational integer for $1 \leq m < n$.

2°) For y real, $0 \le y \le 1$, prove the equalities

$$\sum_{m=1}^{n} \binom{n-1}{m-1} y^{m-1}\, I(m,n) = \int_0^1 (1-x+xy)^{n-1}\,dx = \frac{1}{n}\sum_{m=1}^{n} y^{m-1}.$$

Deduce the relations

$$I(m,n) = \frac{1}{m}\binom{n}{m}, \quad 1 \le m \le n.$$

3°) Prove that d_n divides the product $m\binom{n}{m}$, for $1 \le m \le n$. Deduce that $n\,(2n+1)\binom{2n}{n}$ divides d_{2n+1}. Conclude that $d_{2n+1} \ge n\,4^n$, for every integer $n \ge 1$. Then prove the inequality*

$$d_n \ge 2^n \quad \text{for } n \ge 7.$$

16. Estimations on prime numbers : upper bounds.
If n is a positive integer, we put $\pi(x) = \{p \le n\,;\, p \text{ prime}\}$.

1°) Using the preceding exercise, show that $d_n \le n^{\pi(x)}$, and also

$$\pi(n) \ge \operatorname{Log} 2\, \frac{n}{\operatorname{Log} n}.$$

2°) We want to prove that $\prod_{p\le n} p \le 4^n$ for $n \ge 1$. Use induction on n.

(i) This is true for $1 \le n \le 3$.

(ii) If the result is true for n odd and ≥ 3, then the result is true for $n+1$.

(iii) If $n = 2m+1$, consider $N = \binom{2m+1}{m}$. Show that every prime number p in the interval $m+1 \le p \le 2m+1$ divides N. Deduce that

$$\prod_{m+1}^{2m+1} p \le 4^m,$$

and conclude.

* M. Nair, A new method in elementary prime number theory, *J. London Math. Soc.* (2), 25, 1982, p. 385–391.

Chapter 3

POLYNOMIALS, ALGEBRAIC STUDY

This chapter presents definitions and general algebraic properties of polynomials over arbitrary domains.

1. Definitions and elementary properties

1. First definitions

Let $\{X_1, \ldots, X_n\}$ be a finite set and let A be a unitary commutative ring. The *polynomials* in the *variables* $X_1, \ldots, X_n$ with coefficients in A are the formal sums

$$\sum_{i_1,\ldots,i_n \geq 0} a_{i_1,\ldots,i_n} \, {X_1}^{i_1} \cdots {X_n}^{i_n},$$

where the *coefficients* $a_{i_1,\ldots,i_n}$ belong to A, and only a finite number of the $a_{i_1,\ldots,i_n}$ are different from zero. The set of these polynomials is denoted $A[X_1, \ldots, X_n]$.

A polynomial like $a_{i_1,\ldots,i_n} {X_1}^{i_1} \cdots {X_n}^{i_n}$, with at most one nonzero coefficient, is called a *monomial*. The application $a \longmapsto aX_1^0 \cdots X_n^0$ allows to identify A to a subset of $A[X_1, \ldots X_n]$; this monomial is denoted a, and called a *constant*.

2. Elementary arithmetic operations

Let P and Q be two polynomials of $A[X_1, \ldots X_n]$, which are respectively denoted

$$P = \sum a_{i_1,\ldots,i_n} \, {X_1}^{i_1} \cdots {X_n}^{i_n} \quad \text{and} \quad Q = \sum b_{i_1,\ldots,i_n} \, {X_1}^{i_1} \cdots {X_n}^{i_n},$$

and let λ be an element of A.

We define the following operations

$$P+Q = \sum_{i_1,\dots,i_n\geq 0} (a_{i_1,\dots,i_n} + b_{i_1,\dots,i_n})\, X_1{}^{i_1}\cdots X_n{}^{i_n}\,,$$

$$\lambda\cdot P = \sum_{i_1,\dots,i_n\geq 0} (\lambda\cdot a_{i_1,\dots,i_n})\, X_1{}^{i_1}\cdots X_n{}^{i_n}\,,$$

and

$$P\cdot Q = \sum_{\mathbf{k}}\Big(\sum_{\mathbf{i},\mathbf{j}\,;\,\mathbf{i}+\mathbf{j}=\mathbf{k}} a_{\mathbf{i}}\, b_{\mathbf{j}}\, X^{\mathbf{k}}\Big),$$

where, we have have simplified the notations by setting

$$\mathbf{i} = (i_1,\dots,i_n),\quad \mathbf{j} = (j_1,\dots,j_n),\quad \mathbf{i}+\mathbf{j} = (i_1+j_1,\dots,i_n+j_n),\ \dots$$

and

$$X^{\mathbf{i}} = X_1{}^{i_1}\cdots X_1{}^{i_1},\ \dots$$

With these operations, $A[X_1,\dots,X_n]$ is a commutative A–algebra, and the polynomial $P=1$ is the unit element of this algebra.

If m is an integer, with $1\leq m<n$, we get a natural isomorphism

$$A[X_1,\dots,X_m]\,[X_{m+1},\dots,X_n] \sim A[X_1,\dots,X_n].$$

3. Notions of degree

If P is a nonzero polynomial in X_1, …, X_n (that is, a polynomial for which one of the coefficients, at least, is nonzero),

$$P = \sum a_{\mathbf{i}}\, X^{\mathbf{i}},$$

we define the *total degree* of P by the formula

$$\deg(P) = \max\{i_1+\cdots+i_n\,;\, a_{\mathbf{i}}\neq 0\},$$

whereas the *partial degree* of P *relative to* X_j is defined by

$$\deg_{X_j}(P) = \max\{i_j\,;\, a_{i_1,\dots,i_n}\neq 0\},$$

we, sometimes, also denote it $\deg_j(P)$.

If P and Q are two nonzero polynomials, the following properties hold

- if $\deg(P) \neq \deg(Q)$, then

$$P + Q \neq 0 \quad \text{and} \quad \deg(P+Q) = \max\{\deg(P), \deg(Q)\},$$

- if $\deg(P) = \deg(Q)$ and $P + Q \neq 0$, then

$$\deg(P+Q) \leq \max\{\deg(P), \deg(Q)\},$$

- if $PQ \neq 0$, then

$$\deg(P \cdot Q) \leq \deg(P) + \deg(Q).$$

Remark

It is, sometimes, convenient to define the degree of the zero polynomial by $\deg(0) = -\infty$, with the following conventions

$$n + (-\infty) = (-\infty) + (-\infty) = -\infty$$

for any integer n and $\deg(0) \leq n$ for any integer n.

4. Case of an integral domain

A ring is called an *integral domain* — or an *entire ring* — when the product of any two nonzero elements of A is never zero. When the ring A is an integral domain, we get the following results.

PROPOSITION 3.1. — *If A is an integral domain, then the ring of polynomials $A[X_1, \dots X_n]$ is also an integral domain. Moreover, if P and Q are two nonzero polynomials of $A[X_1, \dots X_n]$, then*

$$\deg(P \cdot Q) = \deg(P) + \deg(Q),$$

and also

$$\deg_j(P \cdot Q) = \deg_j(P) + \deg_j(Q), \quad \textit{for} \quad 1 \leq j \leq n.$$

Proof

First, notice that the relation ρ is defined on $\mathbb{N}^n$ by

$$\mathbf{i}\,\rho\,\mathbf{j} \iff \begin{cases} i_1 + \cdots + i_n < j_1 + \cdots + j_n \\ \text{or} \\ i_1 + \cdots + i_n = j_1 + \cdots + j_n \quad \text{and} \quad \mathbf{i} \leq \mathbf{j}, \end{cases}$$

where the symbol $\leq$ designates the lexicographical order on the set $\mathbb{N}^n$, then ρ is a relation of total order on this set.

Let d and d' be the respective degrees of the polynomials P and Q. Then

$$P = a_{\mathbf{i}}\,X^{\mathbf{i}} + P_1, \quad Q = b_{\mathbf{j}}\,X^{\mathbf{j}} + Q_1, \quad \text{with} \quad a_{\mathbf{i}} \neq 0 \;\text{ and }\; b_{\mathbf{j}} \neq 0,$$

where the polynomial P_1 (and respectively Q_1) only contains coefficients whose indices are smaller than $\mathbf{i}$ for the order ρ (respectively whose indices are smaller than $\mathbf{j}$). Then

$$P \cdot Q = a_{\mathbf{i}}\,b_{\mathbf{j}}\,X^{\mathbf{i}+\mathbf{j}} + R, \quad \text{with} \quad a_{\mathbf{i}}\,b_{\mathbf{j}} \neq 0,$$

where the polynomial R only contains coefficients whose indices are smaller than $\mathbf{i}+\mathbf{j}$ for the order ρ. Therefore, the product $P\,Q$ is nonzero and satisfies

$$\deg(P \cdot Q) = \deg(X^{\mathbf{i}+\mathbf{j}}) = d + d'.$$

The last relation can be demonstrated in a similar but simpler manner, the details are left to the reader. □

2. Euclidean division

1. Monic polynomials

If $P = a_0 + \cdots + a_m X^m$ is a polynomial in one variable with a_m different from zero, then a_m is called the *leading coefficient* of P; sometimes, the leading coefficient of P is denoted $\mathrm{lc}(P)$. When $\mathrm{lc}(P) = 1$, the polynomial P is said to be *monic*.

Remark

If P is monic and Q is nonzero, we always have

$$\deg(P \cdot Q) = \deg(P) + \deg(Q).$$

2. Euclidean division

THEOREM 3.1. — *Let F and G be two polynomials of $A[X]$, where G is monic. Then there exist unique polynomials Q and R such that*

$$F = QG + R \quad \textit{with} \quad R = 0 \quad \textit{or} \quad \deg(R) < \deg(G).$$

Proof

Let us write $d = \deg(F)$ and $d' = \deg(G)$. When, $d < d'$, we get the same type of relation as in theorem 1 with $Q = 0$ and $R = F$.

Let us, now, consider the case where $d \geq d'$. Let a be the leading coefficient of the polynomial F. The polynomial

$$F_1 = F - aX^{d-d'}G$$

has a degree less than d. By induction on d, we can write

$$F_1 = Q_1 G + R_1 \quad \text{with} \quad R_1 = 0 \quad \text{or} \quad \deg(R_1) < d'.$$

Then, a relation similar to the one in Theorem 3.1, holds when we put

$$Q = aX^{d-d'} + Q_1 \quad \text{and} \quad R = R_1 .$$

Let us, now, show the unicity. A relation

$$Q_1 G + R_1 = Q_2 G + R_2 ,$$

with $R_1 = 0$ or $\deg(R_1) < d'$ and with $R_2 = 0$ or $\deg(R_2) < d'$, implies

$$(Q_1 - Q_2)\, G = R_2 - R_1 .$$

Since G is monic, by calculating the degrees of each member, the remark above shows that necessarily $Q_1 = Q_2$. The relation $R_1 = R_2$ follows. □

The polynomial R defined by Theorem 3.1 is called the *rest* of the division of F by G, whereas Q is called the *quotient* of this division.

Theorem 3.1 admits the following generalization, the proof of which is immediate.

COROLLARY. — *Let F and G be two polynomials of $A[X]$, the leading coefficient of the polynomial G being invertible. Then, there are unique polynomials Q and R such that*

$$F = QG + R, \quad \textit{with} \quad R = 0 \quad \textit{or} \quad \deg(R) < \deg(G).$$

3. The case of a field

In the sequel, the letter K designates a (commutative) field. When the ring A is a field K, the statement of Theorem 3.1 and its proof are simplified. In this case, indeed, it is only necessary, after the corollary, to suppose that the polynomial G is nonzero.

The relation $F = QG + R$ is equivalent to a linear system of $d + 1$ equations in $d + 1$ unknowns, which are the coefficients of the unknown polynomials Q and R. The homogeneous associated system is simply $QG + R = 0$. By the argument above, relative to the unicity, the only solution is $Q = R = 0$. Hence the considered system is a system of Cramer* : it admits one and only one solution. However, the demonstration of Theorem 3.1 has the advantage of leading directly to a simple algorithm for the division.

The existence of the euclidean division also leads to the following result (like in the case of the ring $\mathbb{Z}$ of rational integers).

THEOREM 3.2. — *Let I be an ideal of $K[X]$, then there is a polynomial P such that $I = P \cdot K[X]$.*

Proof

If I is reduced to $\{0\}$, then, just take $P = 0$. Suppose that I is nonzero. Let P be an element of I of minimal degree and let F be any element of I. Let R be the rest of the euclidean division of F by P, then the polynomial R belongs to the ideal I. By the definition of P, we necessarily get $R = 0$. Therefore F is a multiple of P. ▯

---

* See the Appendix.

An ideal I of a ring A, which is equal to the set xA of the multiples of some element x, is called a *principal ideal*, such an element x is then called a *generator* of the ideal I. Any other generator of I is equal to ux, where u is any invertible element of A. Here, if $I = P \cdot K[X]$, then any generator of I is equal to λP, where λ is any non zero element of the field K. We shall often write $I = (P)$ instead of $I = P \cdot K[X]$. More generally the ideal generated by a finite family of polynomials $F_1, \ldots, F_r$ will be denoted $(F_1, \ldots, F_r)$.

The fact that any ideal of $K[X]$ is principal, leads to the existence of a g.c.d., with the relation of Bézout, and the existence of an euclidean division leads to an algorithm of Euclid for the computation of the g.c.d. We shall only state the results (except for notations, the demonstrations are the same as in Chapter 1).

THEOREM 3.3. — *Let $F_1, \ldots, F_r$ be polynomials belonging to $K[X]$. Any generator D of the ideal $(F_1, \ldots, F_r)$ is called a g.c.d. of the F_i. There are polynomials $U_1, \ldots, U_r$ such that*

$$U_1F_1 + \cdots + U_rF_r = D.$$

The polynomial D divides each of the F_i and any polynomial P that divides each of the F_i also divides D. However, any other g.c.d. of the F_i is equal to $\lambda \cdot D$, where λ is a nonzero element of the field K. Finally, with the algorithm of Euclid, it is possible to compute a g.c.d. of the F_i and also the associated polynomials U_i.

The polynomials $F_1, \ldots, F_r$ with coefficients in a field K are said to be *relatively prime* when they admit 1 as g.c.d. The theorem implies the following result.

COROLLARY. — *If the polynomials $F_1, \ldots, F_r$ with coefficients in a field K are relatively prime then there exist r polynomials $U_1, \ldots, U_r$ of $K[X]$ such that the relation of Bézout holds :*

$$U_1F_1 + \cdots + U_rF_r = 1.$$

4. Pseudo-division

The euclidean division is extended in the following way.

THEOREM 3.1 bis. — *Let F and G be two polynomials with coefficients in an integral domain A, of respective degrees d and d', and let b be the leading coefficient of the polynomial G. Write $\delta = \max\{d-d'+1, 0\}$. Then there are unique polynomials Q and R, such that*

$$b^\delta F = QG + R, \quad \textit{with} \quad R = 0 \quad \textit{or} \quad \deg(R) < \deg(G).$$

Proof

The demonstration is an immediate generalization of the proof of Theorem 3.1. The replacement of polynomial F by $b^\delta \cdot F$ exactly allows to make the division. ∎

It is, sometimes said that polynomials Q and R of the previous theorem are respectively the *pseudo-quotient* and the *pseudo-rest* of the *pseudo-division* of F by G. The following corollary will be useful later.

COROLLARY. — *Let F and G be two polynomials with coefficients in an integral domain A. If P is a polynomial of $A[X]$ which divides both F and G, then P also divides the pseudo-rest of the division of F by G.*

Proof

This, also, follows from the relation $b^\delta F = QG + R$. ∎

3. The Chinese remainder theorem

We give here a general version of this theorem. The following lemma will be used during the proof of the Chinese remainder theorem.

LEMMA 3.1. — *Let A be a ring (commutative or not) and let I, I_1, ..., I_n be bilateral ideals of A that verify $I + I_k = A$ for $1 \le k \le n$. Then, we get*

$$I + I_1 \cap \cdots \cap I_n = A.$$

Proof

We only have to demonstrate that element 1 belongs to $I+I_1\cap\cdots\cap I_n$. By hypothesis, there are elements $a_k\in I$, $b_k\in I_k$ such that $a_k+b_k=1$, for $1\le k\le n$, and

$$1=\prod_{k=1}^{n}(a_k+b_k)=\sum_{k=1}^{n}a_k\prod_{h\neq k}(a_h+b_h)+b_1\cdots b_n\in I+I_1\cap\cdots\cap I_n .$$

Hence the result. □

THEOREM 3.4. — *Let A be a ring and I, I_1, ..., I_n be bilateral ideals of A that verify $I_1+I_k=A$ if $1\le h<k\le n$. Then*

$$A/(I_1\cap\cdots\cap I_n)\sim\prod_{k=1}^{n}(A/I_k).$$

Proof

We consider the canonical morphism

$$\varphi : A\longrightarrow\prod_{k=1}^{n}(A/I_k).$$

Its kernel is equal to $I_1\cap\cdots\cap I_n$.

We, now, only have to show that the application φ is surjective. We take the elements a_1, ..., a_n of A and we have to show that there exists an element x of A satisfying $x\equiv a_k\pmod{I_k}$ for $1\le k\le n$. We shall give two demonstrations of this fact.

1°) Method of Lagrange

Let k be an index, $1\le k\le n$. By the lemma, we have

$$I_k+\bigcap_{h\neq k}I_h=A;$$

therefore, there exists an element

$$\varepsilon_h\in\prod_{h\neq k}I_h,$$

such that $1-\varepsilon_k \in I_k$. Then $\varepsilon_k \equiv \delta_{hk} \pmod{I_h}$, where (as usual)

$$\delta_{hk} = \begin{cases} 1, & \text{if } h = k \\ 0, & \text{otherwise} \end{cases}$$

is Kronecker's symbol, and the element

$$x = \sum_{k=1}^{n} a_k \varepsilon_k$$

is a solution to the problem.

2°) Method of Newton

We argue by induction on n. The case $n = 1$ is trivial. Let us suppose that n is ≥ 2 and that we have found an element x' of A that satisfies

$$x' \equiv a_k \pmod{I_k} \quad \text{for} \quad 1 \leq k < n\,.$$

Let ε_n be defined as above, it is instantly verified that

$$x = x' + \varepsilon_n\,(a_n - x')$$

is a solution to our problem. □

Corollary. — *Let A be a commutative ring and let $I_1, \ \ldots, \ I_n$ be ideals of this ring. Then, the following assertions are equivalent :*

(i) $A \sim \prod_{k=1}^{n}(A/I_k)$.

(ii) *There exist $e_1, \ \ldots, \ e_n$ in A such that*

$$e_h{}^2 = e_h \quad \textit{and} \quad e_h e_k = 0, \quad 1 \leq h < k \leq n,$$
$$e_1 + \cdots + e_n = 1 \quad \textit{and} \quad I_k = (1-e_k)A, \quad 1 \leq k \leq n,$$

where the notation λA designates the set of the elements λx, when x runs over A.

(iii) *The considered ideals satisfy*

$$I_h + I_k = A \quad \textit{if} \quad 1 \leq h < k \leq n \quad \textit{and} \quad I_1 \cap \cdots \cap I_n = \{0\}.$$

Proof

(i) $\Longrightarrow$ (ii) : If φ is the isomorphism in (i), we only have to write

$$e_k = \varphi^{-1}(\delta_{1k}, \dots, \delta_{nk}).$$

(ii) $\Longrightarrow$ (iii) : An element e of A such that $e^2 = e$ is called an *idempotent* of A. Since $1 - e_h$ belongs to I_h and $e_h = e_h(1 - e_k)$ belongs to I_k, we get $I_h + I_k = A$ if $h \neq k$. Moreover,

$$\prod_{k=1}^{n}(1 - e_k) = 1 - (e_1 + \cdots + e_n) = 0 \implies I_1 \cap \cdots \cap I_n = \{0\}.$$

(iii) $\Longrightarrow$ (i) : Apply the theorem. $\square$

In the case of the polynomials, the Chinese theorem has the following consequence.

COROLLARY. — *Let K be a field and put $A = K[X]$. If polynomials F_1, $\dots$, F_r of A are two-by-two relatively prime, then there exists a natural isomorphism*

$$A/(F_1 \cdots F_r) \sim \prod_{i=1}^{r}\big(A/(F_i)\big).$$

4. Factorization

In this section, A is an integral domain. A nonzero element x of an integral ring B is said to be *irreducible* if, and only if, the quotient ring B/xB is an integral domain. In particular, a polynomial P of $A[X]$ is irreducible if, and only if, the quotient ring $A[X]/(P)$ is a integral domain.

Two elements x and y of B are said to be *associated* if, and only if, there exists an invertible element u of B such that $x = uy$. In particular, two polynomials F and G belonging to $A[X]$ are associated if, and only if, there exists an invertible element u of the ring A such that $F = uG$.

1. *Case of a field*

A polynomial P of $K[X]$ is irreducible if, and only if, the degree of P is positive (that is, P is not a constant) and if P has no divisor Q such that $0 < \deg(Q) < \deg(P)$. Therefore, any polynomial F of $K[X]$ of positive degree admits an irreducible divisor : just take any divisor of F of minimal positive degree.

As in the case of the integers, with the relation of Bézout, we also get the following result.

PROPOSITION 3.2. — *A polynomial P of $K[X]$, of positive degree, is irreducible if, and only if, the quotient ring $K[X]/(P)$ is a field.*

Again, as in the case of integers, and again with the relation of Bézout, we also have a decomposition in products of irreducible elements in $K[X]$. The precise statement is the following.

THEOREM 3.5. — *Any nonzero polynomial F of the ring $K[X]$ admits a decomposition like*

$$F = c\prod_{i=1}^{k} P_i{}^{\alpha_i}, \quad c \in K,$$

where the polynomials P_i are irreducible, two-by-two nonassociated, and where the α_i are positive.

Moreover, if

$$F = c'\prod_{j=1}^{h} Q_j{}^{\beta_j}, \quad c' \in K,$$

where the polynomials Q_j are irreducible, two-by-two nonassociated, and where the β_j are positive, then $h = k$ and there is a permutation σ of the set $\{1, \ldots, k\}$, such that

$$\beta_{\sigma(i)} = \alpha_i \quad \textit{and} \quad Q_{\sigma(i)} \textit{ is associated to } P_i$$

for $i = 1, \ldots, k$.

2. Case of a factorial ring

A ring is said to be *factorial* if it is an integral domain and if each of its nonzero elements admits a unique decomposition into a product of irreducible elements, up to the order of the factors, and up to an invertible factor (as in the statement of Theorem 3.5).

Let A be a factorial ring, K be its field of fractions, and p be an irreducible element of A. Any nonzero element a of K is written, in one and only way :

$$a = p^r b, \quad r \in \mathbb{Z},$$

where b is the quotient of two elements of A, of which none is divisible by p. The integer r is called the *order of* a *at* p, and denoted $\operatorname{ord}_p(a)$. By convention, $\operatorname{ord}_p(0) = +\infty$.

If a and a' are any elements of K, we get

$$\operatorname{ord}_p(aa') = \operatorname{ord}_p(a) + \operatorname{ord}_p(a').$$

If $F = a_0 + \cdots + a_n X^n$ is a nonzero polynomial with coefficients in K, we put

$$\operatorname{ord}_p(F) = \min \left\{ \operatorname{ord}_p(a_i)\,;\, a_i \neq 0 \right\}.$$

Let $\wp$ designate a *system of representatives of the irreducible elements* of A (which means that, on one hand, for every irreducible element q of A, there is an element p of $\wp$ associated to q, and that, on the other hand, the elements of $\wp$ are two-by-two nonassociated). The *content* of F (relative to $\wp$) is defined by the formula

$$\operatorname{cont}_{\wp}(F) = \prod_{p \in \wp} p^{\operatorname{ord}_p(F)}.$$

When there is no possible confusion, we put $\operatorname{cont}(F) = \operatorname{cont}_{\wp}(F)$.

After all these preliminaries, we can demonstrate the following result.

Proposition 3.2 (Gauss lemma). — *Let A be a factorial ring and let K be its field of fractions. If F and G are two nonzero polynomials with coefficients in the field K, then*

$$\operatorname{cont}(F \cdot G) = \operatorname{cont}(F) \cdot \operatorname{cont}(G).$$

Proof

According to the definition of the content of a polynomial, we only have to prove that for every irreducible element p of A, the following relation holds

$$\mathrm{ord}_p(F \cdot G) = \mathrm{ord}_p(F) + \mathrm{ord}_p(G).$$

Let us put $r = \mathrm{ord}_p(F)$ and $s = \mathrm{ord}_p(G)$. Then, there exist two polynomials F_1 and G_1 with coefficients in A such that we have

$$F = up^r F_1, \quad G = vp^s G_1 \quad \text{and} \quad \mathrm{ord}_p(F_1) = \mathrm{ord}_p(G_1) = 0,$$

where u and v are two nonzero elements of K such that

$$\mathrm{ord}_p(u) = \mathrm{ord}_p(v) = 0.$$

By the canonical application $A[X] \longrightarrow (A/pA)[X]$, the two polynomials F_1 and G_1 have respective images f_1 and g_1, which are nonzero. Since the ring A/pA is an integral domain, the product $f_1 g_1$ is nonzero too; therefore, we get $\mathrm{ord}_p(F_1 G_1) = 0$. Hence, the relation

$$\mathrm{ord}_p(F \cdot G) = r + s,$$

which leads to the result. □

Using the Gauss lemma, we can demonstrate the following result.

THEOREM 3.6. — *Let A be a factorial ring and let K be its field of fractions. Then, the ring of polynomials $A[X]$ is also factorial. Moreover, if $\wp$ is a system of representatives of the irreducible elements of A and $\wp'$ is a system of representatives of the irreducible polynomials of the ring $K[X]$, such that each element of $\wp'$ has a content relative to $\wp$ equal to 1, then the union $\wp \cup \wp'$ is a system of representatives of the irreducible elements of the ring of polynomials $A[X]$.*

Example

The ring $\mathbb{Z}[X]$ is factorial. If $\wp$ designates the set of integers and $\wp'$ the set of irreducible polynomials of $\mathbb{Q}[X]$ with integer coefficients, relatively prime, and with a positive leading coefficient, then $\wp \cup \wp'$ is a system of representatives of the irreducible elements of $\mathbb{Z}[X]$.

Proof

Let F be a nonzero polynomial with coefficients in A. In the ring $K[X]$, this polynomial can be decomposed in

$$F = a \prod_{p \in \wp'} P^{n_P}, \quad a \in K,\ a \neq 0,\ n_P \geq 0,$$

and this decomposition is essentially unique.

By the Gauss lemma, a is equal to the content of F (relatively to $\wp$), hence the element a belongs to the ring A and it can be written, essentially in only one way,

$$a = u \prod_{p \in \wp} P^{n_P}, \quad n_P \geq 0,$$

where u is an invertible element of the ring A. Hence the theorem. □

Arguing by induction on the number of variables, we get the following result.

COROLLARY. — *Let A be a factorial ring, then the ring of polynomials in n variables $A[X_1, \ldots, X_n]$ is also a factorial ring. In particular, if K is any field, the ring $K[X_1, \ldots, X_n]$ is a factorial ring for any positive n.*

Example

For all $n \geq 1$, the ring $\mathbb{Z}[X_1, \ldots, X_n]$ is a factorial ring; it certainly is the most important example in computer algebra.

5. Polynomial functions

1. Definition

Let F be a polynomial in $X_1, \ldots, X_n$ with coefficients in A, and let

$$F = \sum a_{i_1,\ldots,i_n} X_1{}^{i_1} \ldots X_n{}^{i_n},$$

and also let B be a ring of which A is a subring. Then, we define the *polynomial function*

$$F^* : B^n \longrightarrow B$$

by

$$F^*(b_1, \ldots, b_n) = \sum a_{i_1,\ldots,i_n} b_1{}^{i_1} \ldots b_n{}^{i_n} .$$

We say that we *substitute* b_i to X_i and, to simplify the notations, we often write $F(b_1, \ldots, b_n)$ instead of $F^*(b_1, \ldots, b_n)$.

The following properties are obvious

$$\begin{aligned} (F + G)^* &= F^* + G^* , \\ (F \cdot G)^* &= F^* \cdot G^* , \\ (a\, F)^* &= aF^* \quad \text{if} \quad a \in A . \end{aligned}$$

Attention : The application $\varphi : F \longrightarrow F^*$ is not always one-to-one : for example, in the ring $A = \mathbb{Z}/2\mathbb{Z}$, for the polynomial $F = X^2 - X$, the associated polynomial function F^* takes the value zero for every element of the ring A. But the application φ is one-to-one when A is an infinite integral domain (see the corollary of Theorem 3.7 below). Thus, when A is an infinite integral domain, without any inconvenience, we can identify a polynomial and the associated polynomial function.

2. Roots of a polynomial

If $F \in A[X]$, an element a of A is called a *root of* F if $F(a) = 0$. We also say that a is a *zero* of the polynomial F.

PROPOSITION 3.3. — *Let F be a polynomial with coefficients in a ring A and let a be an element of A. Then the rest of the euclidean division of F by the polynomial $X - a$ is equal to $F(a)$.*

In particular, a is a root of F if, and only if, $X - a$ divides the polynomial F.

Proof

If the result of the euclidean division of F by $X - a$ is $F = (X - a)\, Q + b$, where b belongs to A, then making the substitution of X by a, we see at once that $b = F(a)$. Then, by definition, a is a root of F if, and only if, $F(a)$ is equal to zero. Hence the second assertion. $\square$

3. Multiplicity of a root

If $F \in A[X]$, an element a of A is called a *root of order* k *of* F if the polynomial F is divisible by $(X-a)^k$ but not by $(X-a)^{k+1}$; if $k \geq 1$, we say that k is the *multiplicity of* the root a. (By Proposition 3.3, this terminology is coherent with that of the previous section.)

The order of a root can be characterized by the following result.

PROPOSITION 3.4. — *Let F be a polynomial with coefficients in A and let a be an element of A. Then a is a root of order k of F if, and only if, there exists a polynomial G with coefficients in A such that*

$$F = (X-a)^k G \quad \textit{and} \quad G(a) \neq 0 .$$

Proof

If a is a root of order k of F, then $(X-a)^k$ divides F. Let G be the quotient. If the element a were a root of G, then $(X-a)^{k+1}$ would divide the polynomial F, but $G(a)$ is nonzero. The reciprocal is obvious. □

COROLLARY. — *Let A be a domain and let $F \in A[X]$ be a nonzero polynomial. Then every root of F in A has an order at most equal to the degree of F.*

More generally, the proposition has the following consequence.

THEOREM 3.7. — *Let A be an integral domain and let $F \in A[X]$ be a nonzero polynomial. Then the sum of the multiplicities of the roots of F that belong to A is at most equal to the degree of F.*

Proof

Consider a nonzero polynomial F with coefficients in A. Let $a_1, \ldots, a_k$ be roots of F of respective multiplicities at least equal to $r_1, \ldots, r_k$. Using k applications of Proposition 3.4, and using the fact that A is an entire ring, we see that the polynomial F can be written as

$$F = (X-a_1)^{r_1} \cdots (X-a_k)^{r_k}\, G(X),$$

where G is a polynomial with coefficients in A. Hence the inequality

$$r_1 + \cdots + r_k \leq \deg(F),$$

which proves the result. □

COROLLARY. — *If A is an infinite integral domain and if $F \in A[X]$ takes the value zero for every a belonging to A, then the polynomial F is equal to zero.*

4. Derivatives and roots

If $F = a_0 + a_1 X + \cdots + a_m X^m$, the (formal) *derivative* of the polynomial F is defined by the formula

$$F' = a_1 + 2a_2 X + \cdots + m a_m X^{m-1}.$$

This derivation has the usual properties :

$$(F+G)' = F' + G',$$
$$(F\,G)' = F'\,G + F\,G',$$
$$(a\,F)' = a\,F' \quad \text{if} \quad a \in A.$$

More generally, we put $F^{(0)} = F$ and $F^{(k)} = (F^{(k-1)})'$ for any positive integer k.

The following result shows the link between the order of F and of its derivative at the same point.

PROPOSITION 3.5. — *If an element a of the ring A is a root of order $k \geq 1$ of a polynomial F of $A[X]$, then we have $F^{(i)}(a) = 0$ for $i = 0,\ 1,\ \ldots,\ k-1$.*

Proof

We only have to prove that $(X-a)^{k-1}$ divides F' and then to argue by induction on the integer k. By hypothesis, the polynomial F can be written $F = (X-a)^k Q$, hence its derivative is equal to

$$F' = (X-a)^{k-1}\left(k\,Q + (X-a)\,Q'\right).$$

Hence the result. $\square$

5. Taylor's formula

Recall that a ring A is said to be of *characteristic* zero when the natural map $\mathbb{Z} \longrightarrow A$, which sends the integer 1 on the unit element of the ring A, is one-to-one.

In the case of one variable, Taylor's formula can be stated as follows.

PROPOSITION 3.6. — *Let K be a field of characteristic zero and let $F \in K[X]$ be a polynomial that can be written $F = a_0 + a_1 X + \cdots + a_n X^n$. If Y is a new variable, then $F(X+Y)$ satisfies the relation, called* Taylor's formula,

$$F(X+Y) = F(X) + Y\,F^{(1)}(X) + \frac{Y^2}{2!}\,F^{(2)}(X) + \cdots + \frac{Y^n}{n!}\,F^{(n)}(X).$$

Proof

Because of the linearity of the applications $F \longmapsto F^{(k)}$, we only have to prove the result when $F = X^m$, for $m \le n$. Therefore,

$$F(X+Y) = (X+Y)^m = \sum_{k=0}^{m} \binom{m}{k}\, Y^k\, X^{m-k},$$

whereas

$$\frac{Y^k}{k!}\,F^{(k)}(X) = Y^k\,\frac{m\,(m-1)\cdots(m-k+1)}{k!}\,X^{m-k} = \binom{m}{k}\, Y^k\, X^{m-k},$$

whence we draw the conclusion. □

In the case of a field of characteristic zero, Taylor's formula shows that the order of a root can be characterized with the derivation. See the following result.

COROLLARY. — *Let F be a polynomial belonging to $K[X]$, where K is a field of characteristic zero, and let a be a root of order k of F, $k \ge 1$, then $F^{(k)}(a)$ is nonzero.*

Proof

Proposition 3.5, joined to Taylor's formula, shows that

$$F(X) = \sum_{m \ge k} \frac{(X-a)^m}{m!}\,F^{(m)}(a).$$

If $F^{(k)}(a) = 0$, then $(X-a)^{k+1}$ divides F and we have a contradiction. Hence, we have the result. □

6. The resultant

Let f and g be two polynomials in one variable and with coefficients in a field K. When f and g have a nontrivial common factor d, say $f = f_1 h$ and $g = g_1 h$, and the equation

$$uf + vg = 0, \quad \text{with} \quad \deg(u) < \deg(g) \quad \text{and} \quad \deg(v) < \deg(f),$$

has the solutions $u = g_1$ and $v = -f_1$.

Conversely, it is obvious that the existence of nonzero solutions to this equation implies the fact that the polynomials f and g admit a common non trivial divisor.

Write

$$f(X) = \sum_{i=0}^{m} a_i X^i \quad \text{and} \quad g(X) = \sum_{j=0}^{n} b_j X^j .$$

Then the preceding equation is equivalent to an homogeneous linear system of $n+m$ relations in $n+m$ unknowns (the coefficients of the polynomials u and v) and we know that this system admits a nontrivial solution if, and only if, its determinant is zero. Here, the determinant is equal to

$$\begin{vmatrix} a_m & a_{m-1} & \dots & a_1 & a_0 & 0 & \dots & 0 \\ 0 & a_m & a_{m-1} & \dots & a_1 & a_0 & \dots & 0 \\ \dots & \dots & \dots & \dots & \dots & \dots & \dots & \dots \\ b_n & b_{n-1} & \dots & b_1 \quad b_0 & 0 & \dots & 0 & \\ \dots & \dots & \dots & \dots & \dots & \dots & \dots & \\ 0 & \dots & \dots & 0 & b_n & b_{n-1} \dots & b_1 & b_0 \end{vmatrix} .$$

More precisely, the coefficients of the matrix associated to this determinant are given by the formulas

$$\begin{aligned} c_{i,j} &= a_{m-j+i}, \qquad \text{for} \quad 1 \le i \le n, \\ c_{n+i,j} &= b_{n-j+i}, \qquad \text{for} \quad 1 \le i \le m, \end{aligned}$$

where

$$a_i = 0 \ \text{ for } \ i \notin \{0, 1, \dots, m\} \quad \text{and} \quad b_j = 0 \ \text{ for } \ j \notin \{0, 1, \dots, n\}.$$

This determinant is called the *resultant* of the polynomials f and g; it is most often written $\text{Res}(f,g)$, or $\text{Res}_X(f,g)$ when it is necessary to precise the variable. More generally, the resultant of f and g is defined by the preceding determinant if the coefficients of these polynomials belong to a ring A. Thus $\text{Res}(f,g)$ is an element of A.

The matrix corresponding to the determinant above is called the *matrix of Sylvester* for the polynomials f and g.

We now prove that in the general case the resultant of f and g is zero if, and only if, the polynomials f and g have a common non trivial factor [cf. property (iv) below].

In fact, the resultant has the following properties :

(i) $\quad \text{Res}(f,0) = 0$.

Proof

This is clear according to the definition. □

(ii) $\quad \text{Res}(g,f) = (-1)^{mn}\ Res(f,g)$.

Proof

This is due to the fact that a determinant changes sign each time we exchange two adjacent rows. □

(iii) If $\deg(f) = m \leq n = \deg(g)$ and if h is the rest of the euclidean division of the polynomial g by the polynomial f, then

$$\text{Res}(f,g) = a_m^{n-m}\ \text{Res}(f,h).$$

Proof

Write $g = fq + h$, and subtract, in each of the m last lines of the matrix associated to the resultant, the vector corresponding to the product polynomial fq (this vector is shifted one position to the right when passing from one line to the next). This is like subtracting, from each of the m last lines of this matrix, a suitable linear combination of its n first lines so that the result is a matrix of the following form

$$Q = \begin{pmatrix} M & M' \\ 0 & N \end{pmatrix},$$

where M is a triangular matrix of size $(n-m) \times (n-m)$, and N is the matrix of Sylvester associated to polynomials f and h. Then the obvious formula

$$\det(Q) = \det(M) \cdot \det(N)$$

implies the result. □

(iv) $\operatorname{Res}(f,g) = 0$ if, and only if, f and g have a common nontrivial factor.

Proof

It is an immediate consequence of properties (ii) and (iii). □

(v) Suppose that ring A is integral and suppose that a_i, $i = 1, \ldots, m$, are the roots of the polynomial f and that b_j, $j = 1, \ldots, n$, are the roots of g (in a suitable extension of ring A), then the following relations hold

$$\begin{aligned}\operatorname{Res}(f,g) &= a_m{}^n \prod_{i=1}^{m} g(\alpha_i) = (-1)^{mn} b_n{}^m \prod_{j=1}^{n} f(\beta_j) \\ &= a_m{}^n\, b_n{}^m \prod_{i=1}^{m}\prod_{j=1}^{n} (\alpha_i - \beta_j)\,.\end{aligned}$$

Proof

It is clear that the second and the third expressions are both equal to the fourth, let us designate by S this quantity. We, now, only have to show that S is nothing else than the resultant of the polynomials f and g.

First, notice that properties (i) to (iii) define the resultant in a unique way : indeed, it is obvious that these three properties allow us to compute the resultant by induction on the minimum of the degrees of the polynomials f and g. We only have to show now that S verifies properties (i) to (iii). For the property (i) this is trivial. For the second it is the result of the equality between the three expressions of S. Finally, the last property is an immediate consequence of the relation $h(\alpha_i) = g(\alpha_i)$, which is true when h is the rest of the euclidean division of the polynomial g by f (use formula $g = qf + h$ at point α_i). □

We also define the *discriminant of a polynomial* f of degree m and roots α_1, α_2, ... by the formula

$$\operatorname{Discr}(f) = {a_m}^{2m-2} \prod_{i=1}^{m} \prod_{j=1\,;\,j\neq i}^{m} (\alpha_i - \alpha_j),$$

where a_m is the leading coefficient of f.

According to property (v) above, $\operatorname{Res}(f, f') = a_n \operatorname{Discr}(f)$, and we can prove that the discriminant of f belongs to the ring A.

7. Companion matrix

Let K be a field and $F \in K[X]$ be a monic polynomial,

$$F(X) = X^m + a_1 X^{m-1} + \cdots + a_{m-1} X + a_m .$$

Consider the quotient ring $A = K[X]/\big(F(X)\big)$ and its canonical basis $B = \{1, X, \ldots, X^{m-1}\}$. The K-linear transformation $T : P \mapsto (X \cdot P) \bmod F$ from A to A, is represented on basis B by the $m \times m$ matrix

$$C = C_F = \begin{pmatrix} 0 & 1 & 0 & \ldots & 0 & 0 \\ 0 & 0 & 1 & \ldots & 0 & 0 \\ \ldots & \ldots & \ldots & \ldots & \ldots & \ldots \\ 0 & 0 & 0 & \ldots & 0 & 1 \\ -a_m & -a_{m-1} & -a_{m-2} & \ldots & -a_2 & -a_1 \end{pmatrix}.$$

This matrix C_F is called the *companion matrix* of the polynomial F. By the definition of the quotient ring A, it is clear that the minimal polynomial of T is equal to F. This proves that F is the minimal polynomial of the matrix C. Hence, the characteristic polynomial of C is also equal to F. In other words, the eigenvalues of the matrix C_F are exactly the roots of F, with the same multiplicities.

Let G be a nonconstant polynomial of $K[X]$. In the sequel we denote $H = H_G = H(G\,;\,F)$ the matrix $G(C_F)$ which represents the transformation $G(T)$ on the basis B.

The eigenvalues of the matrix H are exactly $G(\alpha_1), G(\alpha_2), \ldots, G(\alpha_m)$, where $\alpha_1, \alpha_2, \ldots, \alpha_m$ are the roots of F. This proves that the characteristic polynomial of H is

$$\det(X\,I - H_G) = \prod_{i=1}^{m}\big(X - G(\alpha_i)\big), \tag{1}$$

where I is the identity matrix.

This gives a simple way to produce the polynomial on the right-hand side. More generally, when F is a given polynomial, the roots of which are α_1, α_2, ..., α_m, and when φ is a given rational function, then the *Tschirnhaus transformation* of F by φ is the equation whose roots are $\varphi(\alpha_1)$, $\varphi(\alpha_2)$, ..., $\varphi(\alpha_m)$. If φ is equal to the fraction $P(X)/Q(X)$, we have to suppose that $Q(\alpha_i)$ is not zero for $i = 1, 2, \ldots, m$. In other words, the polynomials F and Q must be coprime. Then, by Bézout's theorem, there exist two polynomials U and V in $K[X]$ such that $UF + VQ = 1$, which implies

$$\varphi(\alpha_i) = P(\alpha_i)\,V(\alpha_i), \quad 1 \le i \le n.$$

This remark shows that we may always suppose that the function φ is a polynomial. And we conclude that any Tschirnhaus transformation can be computed with the companion matrix.

Example

First observe that, in a field of characteristic different from 3, a "general" equation of the third degree $AX^3+BX^2+CX+D=0$ can be transformed into the standard form $X^3+pX+q=0$ by an affine change of variables.

Now, transform the equation $F = X^3 + pX + q$ by the function $\varphi = aX^2 + bX + c$. Here

$$C = \begin{pmatrix} 0 & 1 & 0 \\ 0 & 0 & 1 \\ -q & -p & 0 \end{pmatrix}, \quad C_\varphi = \begin{pmatrix} c & b & a \\ -aq & -ap+c & b \\ -bq & -aq-bp & -ap+c \end{pmatrix}.$$

Hence, the transformed polynomial $\Phi(X) = \det(XI - C_\varphi)$ is equal to

$$\begin{aligned} X^3 &+ (2ap-3c)\,X^2 + \{(c-ap)(3c-ap) + b(3aq+bp)\}X \\ &+ qb^3 + (pa-c)\{2qab - c(pa-c)\} - (qa+pb)(bc+qa^2). \end{aligned}$$

Observe that choosing $a = 3$ and $c = 2p$ (we suppose that the characteristic of the field K is neither 2 nor 3), the coefficient of X^2 in Φ cancels. Then, the coefficient of X in Φ is the quadratic polynomial $pb^2 + 9qb - 3p^2$. The discriminant of this polynomial is $\Delta = 3(27q^2 + 4p^3)$. This shows, that in the field $K' = K[\sqrt{\Delta}]$, the equation $X^3 + pX + q = 0$, can be transformed in the special form $X^3 = A$, for some element A of K'. We have proved that, over a field of characteristic different from 2 and 3, an equation of the third degree can be solved by radicals.

Come back to relation (1). As a special case, we have the important formula

$$\det\big(H(G\,;\,F)\big) = \operatorname{Res}(F, G). \tag{2}$$

Let R be the rest of the euclidean division of G by F. Then $G = Q\,F + R$ for some polynomial Q. Since $F(T) = 0$, we have $G(T) = R(T)$. Which can be written

$$H(G\,;\,F) = H\big((G \bmod F)\,;\,F\big).$$

Put

$$R(X) = r_0 X^p + r_1 X^{p-1} + \cdots + r_p\,, \qquad \text{where } r_0 \neq 0.$$

If $e_i = (0, \ldots, 0, 1, 0, \ldots 0) = \delta_{ij}$ is the ith vector of this basis for $1 \leq i \leq m$, then e_i represents the element X^{i-1} of A. By definition, $e_{i-1}C = e_i$, for $1 < i \leq m$. Thus, the first rows of C, C^2, $\ldots$, C^{m-1} are e_2, e_3, $\ldots$, e_m. Then

$$\begin{aligned} e_1\,H &= r_0\,e_1\,C^p + r_1\,e_1\,C^{p-1} + \cdots + r_p\,e_1\,I \\ &= r_p\,e_1 + r_{p-1}\,e_2 \quad + \cdots + r_0\,e_{p+1}\,, \end{aligned}$$

where I is the identity matrix. This shows that the first row of H is equal to $(r_p, r_{p-1}, \ldots, r_0, 0, \ldots, 0)$.

Let D be the g.c.d of F and G. Let U and V be two polynomials that verify Bézout's relation $UF + VG = D$. Then $V(T)\,G(T) = D(T)$, which shows that $\operatorname{Ker} G(T) \subset \operatorname{Ker} D(T)$. Since the inverse inclusion is trivial, we see that $\operatorname{Ker} G(T) = \operatorname{Ker} D(T)$. Moreover, it is clear that

$$\operatorname{Image} D(T) = \{D, X\,D, \ldots, X^{m-d-1}\,D\},$$

where d is the degree of D. This proves that

$$\operatorname{rank} G(T) = m - d = \deg F - \deg(F, G).$$

The columns of the matrix $H(G\,;\,F)$ are given by the components of the polynomials $G(X) \bmod F$, $XG(X) \bmod F$, ..., $X^{n-1}G(X) \bmod F$. Consider the particular case of the polynomial D, the g.c.d. of F and G, if $D(X) = X^d + c_1\,X^{d-1} + \cdots + c_d$, we get

$$H_D = H(D\,;\,F) = \begin{pmatrix} c_d & 0 & 0 & \ldots \\ c_{d-1} & c_d & 0 & \ldots \\ c_{d-2} & c_{d-1} & c_d & \ldots \\ \ldots & \ldots & \ldots & \ldots \\ 1 & c_1 & c_2 & \ldots \\ 0 & 1 & c_1 & \ldots \\ \ldots & \ldots & \ldots & \ldots \\ 0 & 0 & 0 & \ldots \end{pmatrix}.$$

This shows that the first $m - d$ first columns of the matrix H_D are linearly independent (look at the last $m - d$ components of these vectors). Since the rank of H_D is equal to $m - d$, the last d columns of H_D are linear combinations of the first $m - d$ ones. Moreover, these relations of dependency are of the following type

$$\begin{aligned} \mathcal{C}_{m-d+1} &= c_1\,\mathcal{C}_{m-d} + \cdots, \\ \mathcal{C}_{m-d+2} &= c_2\,\mathcal{C}_{m-d} + \cdots, \\ &\ldots\ldots\ldots \\ \mathcal{C}_{m} &= c_d\,\mathcal{C}_{m-d} + \cdots, \end{aligned}$$

where $\mathcal{C}_i$ represents the i-column of H_D.

Let $G = G_1 D$. Applying the map $G_1(T)$ to these relations gives

$$\mathcal{V}_{m-d+i} = c_i\, \mathcal{V}_{m-d} + \sum_{j=1}^{m-d-1} x_{ij}\, \mathcal{V}_{m-d-j}\,, \quad x_{ij} \in K\,,$$

for $j = 1, 2, \ldots, d$, where $\mathcal{V}_i$ is the the ith column of the matrix H_G.

Suppose $m \geq n$ and come back to the Sylvester matrix of the polynomials F and G, say $S = S(F\,;\,G)$. Decompose this matrix into four blocks

$$S = \begin{pmatrix} T_1 & T_2 \\ T_3 & T_4 \end{pmatrix},$$

where the sizes of the matrices T_i are respectively $n \times n$, $n \times (m-n)$, $(m-n) \times n$, and $(m-n) \times (m-n)$ for $i = 1, 2, 3$, and 4. One verifies† the formula

$$-{T_1}^{-1} T_3\, T_2 + T_4 = J\, H_G\, J\,,$$

where

$$J = \begin{pmatrix} 0 & 0 & \ldots & 0 & 1 \\ 0 & 0 & \ldots & 1 & 0 \\ \ldots & \ldots & \ldots & \ldots & \\ 0 & 1 & \ldots & 0 & 0 \\ 1 & 0 & \ldots & 0 & 0 \end{pmatrix}.$$

This formula implies the relation

$$\begin{pmatrix} I_n & 0 \\ -{T_1}^{-1} T_3 & I_m \end{pmatrix} S = \begin{pmatrix} T_1 & T_2 \\ 0 & JHJ \end{pmatrix},$$

where I_k is the identity matrix of size $k \times k$. Hence the equalities

$$\det S = \det(J\, H\, J) = \det H = \operatorname{Res}(F, G)\,,$$

and the relation

$$\operatorname{rank} S(F\,;\,G) = m + n - d = \deg F + \deg G - \deg(F, G)\,.$$

---

† A reference for this paragraph is the work of S. Barnett .— Greatest Common divisors from generalized Sylvester Resultant Matrices, *Linear and Multilinear Algebra*, 8, 1980, p. 271–279.

Example

Consider the two general monic trinomials of the second degree $F(X) = X^2 + aX + b$ and $G(X) = X^2 + cX + d$. Then

$$C_F = \begin{pmatrix} 0 & 1 \\ -b & -a \end{pmatrix},$$

and

$$H(G\,;\,F) = H(G - F\,;\,F) = \begin{pmatrix} -b+d & -a+c \\ ab - cb & a^2 - ac + d - b \end{pmatrix},$$

so that

$$\operatorname{Res}(F, G) = (a - c)\,(ad - bc) + (d - b)^2.$$

There is another way to compute the resultant, which was found by Cayley. It uses the following polynomial, which was introduced by Bézout, and is called the *bezoutian* of F and G :

$$B(X, Y) = \frac{F(X)\,G(Y) - F(Y)\,G(X)}{X - Y}.$$

Now, we suppose $n = \deg G \le \deg F = m$. We change the notations and put $G(X) = b_0 X^m + b_1 X^{m-1} + \cdots + b_m$, where the first $m - n$ coefficients are equal to zero.

It is clear that the polynomial satisfies $B(X, Y) = B(Y, X)$. Moreover, if we put $B(X, Y) = \sum z_{ij}\, X^{i-1}\, Y^{j-1}$, then z_{ij} is given by

$$z_{ij} = \sum_{h=0}^{\min\{i-1,\,j-1\}} (a_{m-i-j+1+h}\, b_{m-h} - a_{m-h}\, b_{m-i-j+1+h}),$$

[use the formulas $X^p - Y^p = (X - Y) \sum_{\ell=1}^{p-1} X^\ell\, Y^{p-\ell}$]. Put $Z = (z_{ij})$. The previous formula implies the relation

$$e_i\, Z = e_{i+1}\, Z\, C_F + a_{n-i}\, e_n\, Z.$$

Using the fact that the last row $e_n\, Z$ of Z is given by

$$e_n\, Z = (a_0 b_n - a_n b_0, a_0 b_{n-1}, \ldots, a_0 b_1 - a_1 b_0),$$

the previous relation enables us to compute successively the rows $e_{n-1}\,Z$, $e_{n-2}\,Z, \ldots$ and leads also to the relation

$$Z = T_0\,H\,,$$

where T_0 is the triangular matrix given by

$$T_0 = \begin{pmatrix} a_{n-1} & a_{n-2} & \ldots & a_1 & 1 \\ a_{n-2} & a_{n-3} & \ldots & 1 & 0 \\ a_{n-3} & a_{n-4} & \ldots & 0 & 0 \\ \ldots & \ldots & \ldots & \ldots & \ldots \\ a_1 & 1 & \ldots & 0 & 0 \\ 1 & 0 & \ldots & 0 & 0 \end{pmatrix}.$$

Since, clearly, $\det T_0 = (-1)^{m(m-1)/2}$, we get another formula for the resultant of F and G :

$$\mathrm{Res}\,(F,G) = (-1)^{m(m-1)/2}\,\det Z\,.$$

Moreover, we get also the relation

$$\mathrm{rank}\,Z = m - d = \deg F - \deg\,(F,G)\,.$$

Example

Consider the two polynomials $F = a_0X^3 + a_1\,X^2 + a_2\,X + a_3$ and $G = b_0\,X^3 + b_1\,X^2 + b_2\,X + b_3$. Then, an easy computation leads to

$$\begin{aligned} B(X,Y) = {} & (a_1b_0 - a_0b_1)\,X^2Y^2 + (a_2b_0 - a_0b_2)\,(XY^2 + X^2Y) \\ & + (a_3b_0 - a_0b_3)\,(X^2 + Y^2) + \big\{(a_3b_0 - a_0b_3) + (a_2b_1 - a_1b_2)\big\}XY \\ & + (a_3b_1 - a_1b_3)\,(X + Y) + (a_3b_2 - a_2b_3)\,. \end{aligned}$$

Consider the special case where G is the derivative of F. To simplify the notations, write $F = a\,X^3 + b\,X^2 + c\,X + d$. Then the polynomial B reduces to

$$\begin{aligned} -B(X,Y) = {} & ab\,X^2Y^2 + 2ac\,(XY^2 + X^2Y) + 3ad\,(X^2 + Y^2) \\ & + (3ad + bc) + XY + 2bd(X + Y) + c\,. \end{aligned}$$

The computation of the determinant of the associated matrix Z leads to the formula

$$\mathrm{Discr}\,(F) = 18abcd + b^2c^2 - 4b^3d - 4ac^3 - 27a^2d^2.$$

Using the formula between the Sylvester matrix S and the companion matrix H, and the relation $Z = T_0 H$, after some computations, we get¶

$$\begin{pmatrix} I_n & 0 \\ -T_3 & T_1 \end{pmatrix} S = \begin{pmatrix} T_1 & T_2 \\ 0 & ZJ \end{pmatrix}.$$

8. Linear recursive sequences

We have already met one linear recursive sequence : the sequence of Fibonacci. This section is devoted to general recursive sequences, which are very important in many mathematical domains†. To begin this study, we need a lemma in linear algebra.

Let $A = (a_{i,j})_{0 \le i,j \le m}$ be a square matrix with coefficients over a field. Set $A' = (a_{i,j})_{0<i,j<m}$. Let $A_{i,j}$ be the factor of the term $a_{i,j}$ in the expansion of the determinant of A. Set

$$B = \begin{pmatrix} A_{0,0} & 0 & 0 & \dots & 0 & A_{m,0} \\ A_{0,1} & 1 & 0 & \dots & 0 & A_{m,1} \\ A_{0,2} & 0 & 1 & \dots & 0 & A_{m,2} \\ \dots & \dots & \dots & \dots & \dots & \dots \\ A_{0,m-1} & 0 & 0 & \dots & 1 & A_{m,m-1} \\ A_{0,m} & 0 & 0 & \dots & 0 & A_{m,m} \end{pmatrix}.$$

Notice that

$$AB = \begin{pmatrix} D & a_{0,1} & a_{0,2} & \dots & a_{0,m-1} & 0 \\ 0 & a_{1,1} & a_{1,2} & \dots & a_{1,m-1} & 0 \\ 0 & a_{2,1} & a_{2,2} & \dots & a_{2,m-1} & 0 \\ \dots & \dots & \dots & \dots & \dots & \dots \\ 0 & a_{m-1,1} & a_{m-1,2} & \dots & a_{m-1,m-1} & 0 \\ 0 & a_{m,1} & a_{m,2} & \dots & a_{m,m-1} & D \end{pmatrix},$$

where D is the determinant of A.

¶ See S. Barnett *Polynomials and linear control Systems*, Marcel Dekker, New York, 1983, Th. 1.13.

† See, for example, the survey paper by L. Cerlienco, F. Piras and M. Mignotte, "Suites récurrentes linéaires", *L'Enseignement Math.*, 33 (1-2), 1987, p. 67-108.

If D' is the determinant of A', we get the formula

$$D\,D' = A_{0,0}\,A_{m,m} - A_{0,m}\,A_{m,0}\,.$$

When $a = (a_m)_{m\geq 0}$ is a sequence of elements of a field, for each pair (n,k) of nonnegative integers, we associate to a the *Haenkel* matrix $H_n^k = H_n^k(a) = (\alpha_{i,j})$, where $\alpha_{i,j} = a_{n+i+j}$, $0 \leq i,j \leq k$. As a particular case of the previous relation, we get Sylvester's relation

$$\text{(S)} \qquad H_n^k\,H_{n+2}^{k-2} = H_{n+2}^{k-1}\,H_n^{k-1} - (H_{n+1}^{k-1})^2, \quad \text{for } k \geq 1,$$

(we have put $H_n^{-1} = 1$). As a consequence of this relation, we prove the following lemma.

LEMMA 3.2. — *Let (a) be a nonstationary sequence. Suppose that there exist non negative integers k and n_0 such that $H_n^k(a) = 0$, for all $n \geq n_0$. Then, there exist nonnegative integers h and n_1 such that $H_n^h(a) = 0$, for all $n \geq n_1$, and $H_n^{h-1}(a) \neq 0$, for all $n > n_1$.*

Proof

Consider h minimal such that all the determinants H_n^h are zero for n large enough, say for $n \geq n_1$, where n_1 is minimal. Observe that the hypothesis implies that such an h does exist. Notice also that h is positive since $H_n^0 = a_n$.

Now, suppose that $H_m^{h-1} = 0$, $m > n_1$, then (S), applied for $n = m$, gives

$$H_{m+2}^{h-1}\,H_m^{h-1} - (H_{m+1}^{h-1})^2 = 0,$$

and shows that $H_{m+1}^{h-1} = 0$. More generally, the same proof shows that we have $H_{m'}^{h-1} = 0$ for all indices $m' \geq m$, which contradicts the minimality of h. Hence we have the second assertion. □

Now, we are able to prove the following theorem.

THEOREM 3.8. — *Let $f(X) = \sum_{n=-n_0}^{\infty} a_n\,X^{-n-1}$ be a formal nonzero Laurent serie associated to a sequence (a) of elements of a field K, with a_{n_0} different from zero. Then the following conditions are equivalent :*

(i) *The formal series f represents a rational function.*

(ii) *There exists a positive integer h, an integer n_1, and elements $c_1, c_2, \ldots, c_h$ of K such that*

$$a_{n+h} = c_1\, a_{n+h-1} + c_2\, a_{n+h-2} + \cdots + c_h\, a_n\,, \quad \text{for } n \geq n_1\,.$$

(iii) *There exist nonnegative integers h and n_2 as in the above lemma such that $H_n^h(a) = 0$ for $n \geq n_2$.*

(iv) *There exists an integer n_3 such that the Kronecker determinants $K_n(a) = H_n^0(a)$ are zero for $n \geq n_3$.*

(v) *There exist k integers $r_1, r_2, \ldots, r_k$ and elements $\omega_1, \omega_2, \ldots, \omega_k, \alpha_{1,1}, \ldots, \alpha_{1,r_1}, \alpha_{2,1}, \ldots, \alpha_{2,r_2}, \ldots, \alpha_{k,1}, \ldots, \alpha_{k,r_k}$ of some algebraic extension L of K, such that*

$$a_n = \sum_{j=1}^{k} \sum_{i=1}^{r_j} (-1)^i\, \alpha_{i,j}\, {\omega_j}^n \binom{n}{i-1}, \quad \text{for } n \geq 0,$$

where we have put $\binom{m}{0} = 1$ for all m.

Proof

(i) $\Longrightarrow$ (ii) : If $f(X) = P(X)/Q(X)$, where $Q(X) = X^d - c_1\, X^{d-1} - \cdots - c_d$, then $Q(X) \cdot f(X) = P(X)$, which implies the relations

$$a_{n+d} - c_1\, a_{n+d-1} - \cdots - c_d\, a_n = 0, \quad \text{for } n \geq 0.$$

Observe also that $n_0 = \deg(P) - \deg(Q) + 1$.

(ii) $\Longrightarrow$ (iii) : The proof is trivial.

(iii) $\Longrightarrow$ (i) : Let h be minimal such that property (iii) holds. Then, by the lemma, the infinite vectors $V_0 = (a_{n_1}, a_{n_1+1}, \ldots)$, $V_1 = (a_{n_1+1}, a_{n_1+2}, \ldots)$, $\ldots$ and $V_h = (a_{n_1+h-1}, a_{n_1+h}, \ldots)$ generate a vectorial space of dimension h. There is a unique relation $\alpha_h\, V_0 + \cdots + \alpha_1\, V_{h-1} + V_h = 0$ and the product $X^{n_1}\, f(X) \cdot (X^h + \alpha_1\, X^{h-1} + \cdots + \alpha_h)$ is a polynomial.

(ii) $\Longrightarrow$ (iv) : The proof is trivial.

(iv) $\Longrightarrow$ (iii) : Suppose that the Kronecker determinants are all zero for $n \geq n_3$. Then relation (S) shows that $H_n^1 = 0$ for $n \geq n_3$. Then, the same argument shows that $H_n^2 = 0$ for $n \geq n_3$, $H_n^3 = 0$ for $n \geq n_3$, $\ldots$ Thus $H_n^m = 0$ for all m and all $n \geq n_3$. Hence the result.

(i) $\Longrightarrow$ (v) : Suppose that $f(X) = P(X)/Q(X)$. By euclidean division, $f(X) = E(X) + R(X)/Q(X)$ where $E, R \in K[X]$ and $\deg R < \deg Q$. Suppose that $Q(X)$ splits in some extension L of K, say

$$Q(X) = \prod_{i=1}^{k}(X - \omega_i)^{r_i},$$

where the w_i' are distinct and the r_i are positive. Then, using Bézout's relation, it is easy to prove that there exist polynomials U_1, U_2, $\dots$, U_k with coefficients in L such that

$$U_1 \prod_{i\neq 1}(X - \alpha_i)^{r_i} + U_2 \prod_{i\neq 2}(X - \alpha_i)^{r_i} + \cdots + U_k \prod_{i\neq k}(X - \alpha_i)^{r_i} = R,$$

and $\deg(U_j) < r_j$ for $1 \le j \le k$. Thus,

$$f(X) = E(X) + \sum_{i=1}^{k} \frac{U_i(X)}{(X - \omega_i)^{r_i}}.$$

Writing each polynomial U_i on the basis 1, $(X - \omega_i)$, $(X - \omega_i)^2$, $\dots$ we get the formula

$$f(X) = E(X) + \sum_{i=1}^{k} \sum_{j=1}^{r_i} \frac{\beta_{i,j}}{(X - \omega_i)^j},$$

for some elements $\beta_{i,j}$ of L. Then, it is easy to prove that, for any positive integer j, we have

$$\frac{1}{(1-T)^j} = \sum_{n=0}^{\infty} \binom{n+j-1}{j-1} T^n,$$

where we have put $\binom{n}{0} = 1$ for all n [use induction on j]. Using this relation several times, we get

$$f(X) = E(X) + \sum_{n=0}^{\infty} \left\{ \sum_{i=1}^{k} \sum_{j=1}^{r_i} \beta_{i,j}\, \omega_i^{n+1-j} \binom{n}{i-1} \right\} X^{-n-1}.$$

This formula obviously implies relation (v).

(v) $\Longrightarrow$ (ii) : Suppose now that property (v) holds. Let Δ be the shift-operator which transforms a sequence $(u_n)_{n\geq 0}$ into the sequence $(u_{n+1})_{n\geq 0}$. We shall show that the sequence

$$(\Delta - \omega_1\, Id)^{r_1} \cdots (\Delta - \omega_k\, Id)^{r_d} \cdot (a_n)$$

is ultimately zero.

Since the operators $(\Delta - \omega_i\, Id)$ and $(\Delta - \omega_j\, Id)$ commute, it suffices to prove that the sequences

$$(\Delta - \omega\, Id)^{h'} \cdot \left\{\binom{n}{h-1} \omega^n\right\}$$

are zero for any choice of h and h' such that $h' \geq h \geq 1$.

We argue by induction on h'. First, the result is clear when $h' = 1$. Now, suppose that $h' > 1$ and that the result has been proved up to $h'-1$. Then, the relation

$$\begin{aligned}(\Delta - Id) \cdot \left\{\binom{n}{h-1} \omega^n\right\} &= \left(\left\{\binom{n+1}{h-1} - \binom{n-1}{h-1}\right\}\omega^{n+1}\right) \\ &= \left\{\binom{n}{h-2}\omega^{n+1}\right\} = \omega \cdot \left\{\binom{n}{h-2}\omega^n\right\}\end{aligned}$$

allows to use the induction hypothesis, and the result follows. $\square$

Exercises

1. A polynomial of $A[X_1, \ldots, X_n]$ is called *homogeneous* if it is given by some formula like

$$\sum a_{i_1,\ldots,i_n} X_1{}^{i_1} \cdots X_n{}^{i_n} \quad \text{where} \quad i_1 + \cdots + i_n = \text{constant}.$$

Suppose that the ring of base A is a domain and that u, v, w are polynomials of $A[X_1, \ldots, X_n]$ satisfying $u = vw$. Demonstrate that if u is homogeneous then the polynomials v and w are also homogeneous.

2. The partial derivatives of a polynomial are defined in a usual way. Demonstrate the following equivalence (due to Euler)

$$\sum_{i=1}^{n} \frac{\partial u}{\partial X_i}(X) \cdot X_i = r \cdot u(X) \iff u \text{ is homogeneous of degree } r,$$

where $u \in A[X_1, \ldots, X_n]$, and A is a ring of characteristic 0.

3. Let $F \in A[X_1, \ldots, X_n]$ be a non zero polynomial, where A is a domain. We suppose that, for each $i = 1, \ldots, n$, H_i is some subset of the ring A that satisfies $\mathrm{Card}(H_i) > \deg_i(F)$. Demonstrate that there exists at least a point $\mathbf{a} \in H_1 \times \cdots \times \cdot H_n$ such that $F(\mathbf{a})$ is nonzero.

4. Prove that the polynomial $X^2 + 1$ has infinitely many zeros in the field of the quaternions over $\mathbb{C}$.

5. Let K be a field, and $x_1, \ldots, x_n$ distinct elements of K. We put $w(X) = (X - x_1) \cdots (X - x_n)$. Prove that the polynomial

$$F(X) = w(X) \cdot \sum_{i=1}^{n} \frac{y_i}{w'(x)\,(X - x_i)}$$

is the unique polynomial of $K[X]$ of degree $< n$ such that

$$F(x_i) = y_i \quad \text{for} \quad i = 1, \ldots, n;$$

this polynomial is called the *Lagrange polynomial* (associated to the points x_i and to the values y_i).

Deduce that if $G \in K[X]$ is a polynomial of degree $< n-1$ then

$$\sum_{i=1}^{n} \frac{G(x_i)}{w'(x_i)} = 0.$$

[Consider the polynomial $XG(X)$.]

6. Let $x_1, \ldots, x_r$ be distinct elements of a field K of characteristic zero (this means that the natural map $\mathbb{Z} \to K$ is one-to-one), and $n_1, \ldots, n_r$ be positive integers whose sum is equal to n. Let $y_{i,j}$ be any elements of the field K, for $1 \leq j \leq n_i$, $1 \leq i \leq r$. Prove that there exists a unique polynomial $F \in K[X]$ such that

$$\deg(F) < n \quad \text{and} \quad F^{(j-1)}(x_i) = y_{i,j} \quad \text{for} \quad 1 \leq j \leq n_i, 1 \leq i \leq r;$$

This result is due to Hermite and this polynomial is called the *Hermite polynomial.*

7. Let A be an integral domain and let n be an integer ≥ 1. Demonstrate that a nonzero polynomial $P(X_1, \ldots, X_n)$ with coefficients in A is invertible in $A[X_1, \ldots, X_n]$ if, and only if, its degree is zero and it is invertible in the ring A.

8. Let A be an integral domain that is not a field. Demonstrate that the ring $A[X]$ is not principal.

[Recall that a ring is said to be *principal* when it is a commutative ring whose ideals are all principal : in other words, for every ideal I of A, there exists an element a of A for which the ideal I is equal to the set $(a) = \{ax\,;\, x \in A\}$.]

9. Let A be a principal ring and let $I_1 \subset \cdots \subset I_n \subset \cdots$ be an increasing chain of ideals of A. Prove that this chain is stationary (that is, there exists some index k such that $I_m = I_k$ for $m \geq k$).

10. What is the number of roots of the polynomial $X^3 - X$ in the quotient ring $\mathbb{Z}/6\mathbb{Z}$?

11. Let A be a factorial ring whose field of fractions is K.

1°) Let $F \in A[X]$ be a polynomial of positive degree, irreducible in the ring $A[X]$.

(i) Prove that F is also irreducible in $K[X]$.

(ii) Conclude that if F is a monic polynomial of $A[X]$ and if a is a root of F belonging to the field K, then this root a in fact belongs to the ring A.

2°) Suppose that F and G belong to $A[X]$ and have a nontrivial common factor in $K[X]$. Show that these polynomials also have a nontrivial common factor in $A[X]$.

Deduce the following result : Let L be any field. Suppose that F and G belong to $L[X,Y]$ and have none nontrivial common factor in $L[X,Y]$. Then, these polynomials also have no nontrivial common factor in the ring $L(X)[Y]$.

12. Let A be a ring. For $a \in A$ and $F \in A[X]$, we designate by $\operatorname{ord}(F,a)$ the order of a as a root of F [obviously $\operatorname{ord}(F,a) \geq 0$]. If $G \in A[X]$, demonstrate the following relations

$$\operatorname{ord}(F+G,a) \geq \inf\{\operatorname{ord}(F,a), \operatorname{ord}(G,a)\},$$

$$\operatorname{ord}(F,a) \neq \operatorname{ord}(G,a) \Longrightarrow \operatorname{ord}(F+G,a) = \inf\{\operatorname{ord}(F,a), \operatorname{ord}(G,a)\},$$

$$\operatorname{ord}(FG,a) \geq \operatorname{ord}(F,a) + \operatorname{ord}(G,a).$$

Prove also that equality holds in the last relation when A is an integral domain.

13. Let n be an integer, $n \geq 1$. Let $X_1, \ldots, X_n, Y_1, \ldots, Y_n$ be distinct variables. We put $\mathbf{X} = (X_1, \ldots, X_n)$, $\mathbf{Y} = (Y_1, \ldots, Y_n)$ and we define the sum $\mathbf{X}+\mathbf{Y} = (X_1+Y_1, \ldots, X_n+Y_n)$.

Let $u \in A[\mathbf{X}]$. We define the polynomials $\Delta_\nu u$, in $A[\mathbf{X}]$, by the formula

$$u(\mathbf{X}+\mathbf{Y}) = \sum_\nu (\Delta_\nu)(\mathbf{X}) \cdot \mathbf{Y}^\nu .$$

1°) Prove the formulas

$$u(\mathbf{X}) = \sum_{\nu} (\Delta_\nu u)\,(\mathbf{0})\,\mathbf{X}^\nu\,,$$

$$\Delta_\sigma(uv) = \sum_{\nu+\rho=\sigma} (\Delta_\nu u)\cdot(\Delta_\rho v)\,, \quad \text{if } v \in A[\mathbf{X}]\,.$$

(The letters ν, ρ, σ designate multiindices belonging to $\mathbb{N}^n$.)

2°) Demonstrate the relation

$$(\mathbf{X}+\mathbf{Y})^\nu = \sum_{\nu+\rho=\sigma} \frac{(\rho+\sigma)!}{\rho!\,\sigma!}\,\mathbf{X}^\rho\,\mathbf{Y}^\sigma\,.$$

(Where $\rho! = \rho_1!\cdots\rho_n!$ when $\rho = (\rho_1,\dots,\rho_n)$ and the analogues ...)

Let Z_1, ..., Z_n be new letters, considering the possible developments of the expression $u(\mathbf{X}+\mathbf{Y}+\mathbf{Z})$ deduce the relation

$$\Delta_\rho\,\Delta_\sigma(u) = \frac{(\rho+\sigma)!}{\rho!\sigma!}\,\Delta_{\rho+\sigma}(u)\,.$$

3°) If $\nu = (\nu_1,\dots,\nu_n)$, and if we put

$$D^\nu = \frac{\partial^{\nu_1}}{\partial X_1^{\nu_1}}\cdots\frac{\partial^{\nu_n}}{\partial X_n^{\nu_n}}\,,$$

demonstrate the formula

$$D^\nu\,u = \nu!\,\Delta_\nu\,u\,.$$

4°) If any integer n is invertible in A, demonstrate Taylor's formulas in several variables :

$$u(\mathbf{X}+\mathbf{Y}) = \sum_{\nu} \frac{1}{\nu!}\,(D^\nu u)\,(\mathbf{X})\,\mathbf{Y}^\nu\,,$$

and

$$u(\mathbf{X}) = \sum_{\nu} \frac{1}{\nu!}\,D^\nu u(\mathbf{a})\cdot(\mathbf{X}-\mathbf{a})^\nu\,,$$

when $\mathbf{a}$ belongs to A^n.

14. Demonstrate the formula of Leibnitz

$$(f \cdot g)^{(k)} = \sum_{i=0}^{k} \binom{k}{i} f^{(i)} \cdot g^{(k-i)} \quad \text{for} \quad f, g \in A[X].$$

15. Let A be an infinite integral domain. Let F, G_1, $\ldots$, G_n be polynomials belonging to $A[X_1, \ldots, X_n]$, with $G_i \neq 0$ for all i. We suppose that

$$\forall \mathbf{a} \in A^n, \Big(G_1(\mathbf{a}) \neq 0 \text{ and } \ldots \text{ and } G_n(\mathbf{a}) \neq 0\Big) \Longrightarrow F(\mathbf{a}) = 0.$$

Prove that the polynomial F is zero.

16. We call s_0, s_1, $\ldots$, s_n the elementary symmetrical functions in the variables X_1, $\ldots$, X_n. That is

$$s_0 = 1, \quad s_1 = \sum_i X_i, \quad s_2 = \sum_{i<j} X_i X_j, \quad \ldots, s_n = \prod_i X_i.$$

Demonstrate the formulas

$$\prod_{i=1}^{n}(1 + TX_i) = \sum_{k=0}^{n} s_k T^k, \quad \prod_{i=1}^{n}(X - X_i) = \sum_{k=0}^{n}(-1)^{n-k} s_{n-k} X^k.$$

17. Let $f \in K(X_1, \ldots, X_n)$ be a symmetrical rational function. Demonstrate that there exist two symmetrical polynomials u and v, which belong to the ring $K[X_1, \ldots, X_n]$ and such that $f = u/v$.

18. We put

$$p_k = \sum_{i=1}^{n} X_i{}^k \quad \text{for} \quad k \geq 1.$$

Demonstrate the formulas of Newton

$$p_k = \sum_{i=1}^{k-1} s_i \, p_{k-i} + (-1)^{k-1} k \, p_k,$$

where the s_i are defined as in Exercise 16.

19. Let A be an integral domain. We suppose that a_1, ..., a_n, and b_1, ..., b_n are elements of A that satisfy

$$t_k(a_1,\ldots,a_n) = t_k(b_1,\ldots,b_n) \quad \text{for} \quad 1 \le k \le n,$$

where t_k designates symmetrical polynomial s_k (respectively p_k), as in the previous exercise. Demonstrate that there exists some permutation σ of $\{1,\ldots,n\}$ such that $b_{\sigma(i)} = a_i$ for the indices $i = 1, \ldots, n$.

20. Let K be an infinite field and let E be a vectorial space of finite dimension over K. Suppose that V_1, ..., V_n constitute a finite family of proper subspaces of E (in other words, $V_i \underset{\neq}{\subset} E$ for $i = 1, \ldots, n$). Prove that the union of the V_i is not equal to E.

[Let d be the dimension of E over K. Choose any basis of E. Then consider nonzero polynomials $P_i[X_1,\ldots,X_n]$, for $1 \le i \le n$, such that each function P_i is identically zero on the subspace E_i. To conclude, consider the product $P_1 \cdots P_n$.]

21. Put $A = Z[\sqrt{-5}]$. Consider the polynomial $f(X) = 3X^2 + 4X + 3$.

1°) Prove that the polynomial f is irreducible in $A[X]$.

2°) Demonstrate that, however, f is not irreducible in the field of fractions of the ring A.

[Note the relation $f(X) = (3X + 2 + \sqrt{-5})(3X + 2 - \sqrt{-5})/3$.]

22. Let A be a commutative ring and let f be a nonzero polynomial of $A[X]$ which divides zero in $A[X]$. Demonstrate the theorem of McCoy :

There exists an element c of A such that $c \cdot f = 0$.

[Consider a nonzero polynomial g of minimal degree such that $g \cdot f = 0$.]

23. Prove that if f is a univariate polynomial with coefficients in an integral domain then

$$f \text{ quadratfrei} \iff \operatorname{Discr}(f) \neq 0.$$

24. Let

$$f(X) = a_0X^2 + a_1X + a_0 \quad \text{and} \quad g(X) = b_0X^2 + b_1X + b_0$$

be two trinomials.

1°) Demonstrate the formula

$$\operatorname{Res}(f,g) = (b_0 - a_0)^2 - (b_1 - a_1)(a_1b_0 - a_0b_1).$$

2°) When f and g have real coefficients and α_1, α_2 are two real roots of the polynomial f (which satisfy $\alpha_1 < \alpha_2$) then $\operatorname{Res}(f,g) < 0$ if, and only if, the polynomial g has two real roots β_1 and β_2 real and such that $\alpha_1 < \beta_1 < \alpha_1 < \beta_2$.

25. Let $P(X)$ be a nonconstant polynomial with coefficients in a field and with the roots $\alpha_1, \ldots, \alpha_n$.

1°) Prove that the roots of the polynomial

$$R(Y) = \operatorname{Res}_X\big(P(X+Y), P(-X)\big)$$

are the elements $\alpha_i + \alpha_j$, $1 \le i, j \le n$.

2°) We suppose that all the roots of P are nonzero. Prove that the roots of polynomial

$$S(Y) = \operatorname{Res}_X\big(P(XY), P(X)\big)$$

are the elements α_i/α_j, $1 \le i, j \le n$.

3°) By analogy, determine a formula for a polynomial whose roots are the elements $\alpha_i \cdot \alpha_j$, for $1 \le i, j \le n$.

26. Let $F(X) = P(X)/Q(X)$ be a rational fraction with coefficients in a field K, which is written in simplified form (that is, where P and Q are relatively prime). Let $R(X)$ be a polynomial with coefficients in K and whose roots are $\alpha_1, \ldots, \alpha_n$. Suppose that

$$Q(\alpha_i) \ne 0 \quad \text{for} \quad i = 1, 2, \ldots, n.$$

Demonstrate that there exists a polynomial $V(X)$ of degree smaller than the degree of R such that

$$F(\alpha_i) = P(\alpha_i) \cdot V(\alpha_i) \quad \text{for} \quad i = 1, 2, \ldots, n.$$

27. Let K be a field and consider $f(X)$ and $h(X)$, two polynomials with coefficients in K. Let $\alpha_1, \dots, \alpha_n$ be the roots of f.

1°) Prove that the roots of the polynomial

$$g(Y) = \operatorname{Res}_X\Big(f(X), Y - h(X)\Big)$$

are the elements $h(\alpha_i)$ for $i = 1, 2, \dots, n$.

2°) When $h(X) = X^k$, $k \geq 2$, demonstrate the formula

$$g(Y) = (-1)^{kn} \prod_{i=0}^{n-1} f(\zeta^i Z),$$

where ζ is a primitive kth root of unity (that is, $\zeta^k = 1$ but $\zeta^j \neq 1$ for the integers $j = 1, \dots, k-1$) and where Z satisfies $Z^k = Y$. Deduce the method of Graeffe (case $k = 2$).

28. Let R be a ring, A a subring of R, and x an element of R.

Prove that the following properties are equivalent :

(i) There exist elements $a_1, \dots, a_n$ of A such that

$$x^n + a_1 x^{n-1} + \cdots + a_{n-1}x + a_n = 0,$$

(then x is said to be an *integer* over A).

(ii) The ring $A[x]$ is A-module of finite type.

(iii) There exists a subring B of R which contains A and which is a A-module of finite type.

[In order to demonstrate that (iii) implies (i) we can take a system of generators $y_1, \dots, y_k$ of the A-module B and write

$$x\, y_i = \sum_{j=1}^{k} a_{i\,j}\, y_j \qquad \text{for} \quad 1 \leq i \leq k.$$

Deduce that the y_i are solutions of a homogeneous system of k equations. Let Δ be the determinant of this system. Prove that $\Delta = 0$ and that there exists a sequence of n elements $b_1, \dots, b_n$ of A such that this determinant satisfies $\Delta = x^n + b_1 x^{n-1} + \cdots + b_n$.]

2°) Deduce that the set A' of the elements of R that are integers over A is a subring of R. Give another demonstration of this result using Exercise 25 above.

3°) If R is an integral domain, prove that A' is a field if, and only if, A is also a field.

29. Let

$$f(X) = a_n X^n + \cdots + a_0$$

be a polynomial with coefficients in a field and let $\alpha_1, \ldots, \alpha_n$ be its roots.

1°) Prove that the discriminant Δ of this polynomial satisfies

$$\Delta = (-1)^{n(n-1)/2} \Big(\prod_{1 \le i < j \le n} (\alpha_i - \alpha_j)^2 \Big) a_n^{2n-2}.$$

2°) For any nonnegative k integer we define the kth Newton sum

$$S_k = \sum_{i=1}^{n} {\alpha_i}^k .$$

Let D be the determinant of the matrix $A = (a_{i\,j})_{1 \le i,j \le n}$ where $a_{i\,j} = {\alpha_i}^{j-1}$. Demonstrate the formula of Vandermonde

$$D = \prod_{1 \le i < j \le n} (\alpha_i - \alpha_j).$$

Deduce a relation between Δ and D^2.

Demonstrate, now, the relations

$$D^2 = \det(A)\ \det({}^t A) = \det(S)$$

where

$$S = (s_{i\,j})_{1 \le i,j \le n}, \quad \text{with} \quad s_{i\,j} = S_{i+j-2}.$$

30. Let A and B be two polynomials with coefficients in a ring R and let a be an element of this ring. Using elementary manipulations on the matrix of Sylvester, demonstrate the formula

$$\operatorname{Res}\Big((X-a)\,A(X), B(X)\Big) = B(a)\cdot \operatorname{Res}\Big(A(X), B(X)\Big).$$

Deduce formulas (v) of Section 3.5, and also the relation

$$\operatorname{Res}\Big(A(X), B(X)\,C(X)\Big) = \operatorname{Res}\Big(A(X), B(X)\Big)\cdot \operatorname{Res}\Big(A(X), C(X)\Big)$$

when C is a polynomial with coefficients in R.

31. When f and g are polynomials with coefficient in a field, we have seen that these polynomials have a common nontrivial factor if, and only if, there exist two polynomials u and v such that

$$uf + vg = 0, \quad \text{with} \quad \deg(u) < \deg(g) \text{ and } \deg(v) < \deg(f).$$

Generalize this result in case of polynomials with coefficients in a factorial ring.

32. Let A be a ring and let f and g be two homogeneous polynomials belonging to $A[X_1, \ldots, X_s, Y]$, of respective degrees m and n, and whose degrees in Y are also respectively equal to m and n. Prove that the resultant in Y of these polynomials is either zero or an homogeneous polynomial of degree mn in $A[X_1, \ldots, X_s, Y]$.

[Multiply by t each of the indeterminates $X_1, \ldots, X_s$ and consider the matrix of Sylvester of the two polynomials obtained by this transformation on f and g.

By multiplying the ith line of this matrix by t^{n-i+1} for $1 \le i \le n$, and the $(n+j)$th line by t^{m-j+1} for indices $j = 1, \ldots, m$. Prove that the resultant $R = \operatorname{Res}_Y(f, g)$ satisfies the formula

$$t^p R(tX_1, \ldots, tX_s) = t^p R(X_1, \ldots, X_s),$$

where

$$p = n(n+1)/2 + m(m+1)/2 \quad \text{and} \quad q = (m+n)(m+n+1)/2.$$

Conclude.]

Deduce a demonstration of property (v) of the resultant (see Section 3.5).

33. The aim of this exercise is to prove that the formula giving the resultant is irreducible. More precisely, if $u_0, \ldots, u_m$ and $v_0, \ldots, v_n$ are $2n$ indeterminates and if $f(X) = u_0X^m + \ldots + u_{m-1}X + u_m$ and $g(X) = v_0X^n + \ldots + v_{n-1}X + v_n$ are two polynomials, with coefficients in a field K, whose resultant is

$$R(\mathbf{u}, \mathbf{v}) = \operatorname{Res}_X(f, g)$$

then the polynomial $R(\mathbf{u}, \mathbf{v})$ is irreducible in the ring $K[\mathbf{u}, \mathbf{v}]$. We use induction on the integer $s = m + n$.

1°) Consider the case $s = 2$.

2°) Suppose the result true to order $s - 1$. Consider the polynomial

$$Q = R(0, u_1, \ldots, u_m, v_0, \ldots, v_n).$$

(i) Demonstrate the formula

$$Q = (-1)^n v_0 R_0(u_1, \ldots, u_m, v_0, \ldots, v_n),$$

where R_0 is the resultant (with respect to X) of two polynomials in $s + 1$ variables.

(ii) Deduce that if R is reducible then it can only admit one decomposition such that $R = ST$ where

$$S = au_0 + bv_0, \quad a, b \in K.$$

However, prove that a and b are nonzero.

[Remark that R contains the monomial $u_0^n v_n^m$ ...]

(iii) Demonstrate that a decomposition of this type is impossible and conclude.

[Such a decomposition would imply the fact that any two polynomials f and g, as above, whose coefficients u_0 and v_0 satisfy $au_0 + bv_0 = 0$ always have a common root ...

Another method : use the fact that the polynomials S and T must be homogeneous with respect to each of the families of variables $u_0, \ldots, u_m$ and $v_0, \ldots, v_n$, which implies $ab = 0$ ...]

34. Let f and g be two polynomials with coefficients in a ring A. Demonstrate that there exist two polynomials u and v with coefficients in A such that

$$\operatorname{Res}(f,g) = u(X)\, f(X) + v(X)\, g(X),$$

with $\deg(u) < \deg(g)$ and $\deg(v) < \deg(f)$.

[Let $C_1, \ldots, C_{m+n}$ be the column vectors of the matrix of Sylvester of the polynomials f and g. Replace the last column C_{m+n} by the linear combination $C_{m+n} + XC_{m+n-1} + X^2C_{m+n-1} + \cdots + X^{m+n-1}C_1 \ldots$]

Deduce the property (iv) of the resultant (see Section 3.5).

35. Let F and G be two polynomials with coefficients in a field K, of respective degrees m and n. Suppose that there exists a polynomial D of degree d such that

$$F(X) = D(X)F^*(X), \quad G(X) = D(X)G^*(X), \quad \text{with } F^*, G^* \in K[X].$$

Let A be the matrix of the coefficients of the polynomials $X^{n-1}F(X)$, $X^{n-2}F(X)$, ..., $XF(X)$, $F(X)$, $X^{m-1}G(X)$, $X^{m-2}F(X)$, ..., $XC(X)$, $G(X)$, on the usual basis $(X^{m+n-1}, \ldots, X, 1)$.

For $0 \leq k \leq \min\{m,n\}$ we designate by A_k the matrix obtained from A by suppression of the rows (lines and columns) of indices $< k$ and those of indices $> m+n-k$. (Hence A_0 is equal to A, A_1 is the matrix A without all its elements which are on the "border" ...)

1°) Prove that the polynomials F^* and G^* have a common nontrivial factor if, and only if, there are polynomials U and V with coefficients in K such that

$$UF + VG = 0 \quad \text{with } \deg(U) \leq n-d-1 \text{ and } \deg(V) \leq m-d-1.$$

2°) We consider D fixed but F^* and G^* variable. Demonstrate that

$$R^* = \operatorname{Res}_X(F^*, G^*)$$

is an homogeneous polynomial of degree $m+n-2d$ with respect to the set of the indeterminates $a_0, \dots, a_{m-d}, b_0 \dots, b_{n-d}$. Prove that the determinant S of the matrix A_d is also an homogeneous polynomial of degree $m+n-2d$ with respect to $a_0, \dots, a_{m-d}, b_0 \dots, b_{n-d}$.

Deduce that R^* and S are equal up to a multiplicative nonzero constant factor.

[Use the irreductibility of the resultant : Exercise 33. See also the book by Lelong-Ferrand and Arnaudiès¶.]

36. Let P be a polynomial with integer coefficients whose factorization over the complex numbers is

$$P(X) = a\,(X - \alpha_1)\,(X - \alpha_2) \cdots (X - \alpha_n).$$

Let r be an integer, with $1 \le r \le n$. Show that the number

$$a\,\alpha_1\,\alpha_2\,\cdots\,\alpha_r$$

is an algebraic integer (see Exercise 28 for the definition).

[Considering the computation of the euclidean division, prove that if $Q(X)$ is a polynomial with coefficients that are algebraic integers and for which α is a root, then the polynomial $Q(X)/(X - \alpha)$ is still with integer algebraic coefficients. The result is an obvious consequence of this fact, since the constant term of the polynomial

$$P(X)/((X - \alpha_{r+1}) \dots (X - \alpha_n))$$

is the number $a\alpha_1\alpha_2 \cdots \alpha_r$.]

¶ J. Lelong-Ferrand and J.M. Arnaudiès, *Cours de Mathematiques*, tome 1, Algèbre, Dunod, Paris, 1971.

37. Let K be a field, $P \in K[X]$ a polynomial of positive degree n and m an integer satisfying $0 \le m \le n$. Let $a_1, \dots, a_m$ be m distinct elements of K. Prove that the polynomials $P(X)$, $P(X+a_1)$, ..., $P(X+a_m)$, 1, X, ..., X^{n-m-1} are linearly independent over K.

[Admitting that we have a linear relation with coefficients in K like

$$u_0P(X)+u_1P(X+a_1)+\dots+u_mP(X+a_m)+v_0+v_1X+\dots+u_{m'}X^{m'}=0,$$

where $m' = n - m + 1$, prove that the sum of the u_i is zero. Hence we can rewrite the previous relation as

$$u_1(P(X+a_1)-P(X))+\dots+u_m(P(X+a_m)-P(X))+v_0 \\ +v_1X+\dots+u_{m'}X^{m'}=0,$$

Deduce that it is possible to argue by induction on the degree n of the polynomial P ...]

38. Let K be a field and $P \in K[X]$ be an irreducible polynomial of degree n.

1°) Demonstrate that there exists a field L admitting K as a subfield and in which the polynomial P has a root.

[Consider the quotient ring $K[X]/(P)$ ...]

Moreover, prove that the *degree* of L over K (that is the dimension of L as a vectorial space over K) is equal to n and that L is unique up to an isomorphism.

2°) Deduce that there exists a field L' admitting K as a subfield and in which the polynomial P is completely factorized in factors of the first degree. Moreover, prove that the degree of L' on K is a divisor of $n!$ and that L' is unique up to an isomorphism. The field L' is called the *splitting field*, or the *field of the roots*, of the polynomial P.

39. Demonstrate the following result due to D. Masser : let K be an infinite field, let Q be a polynomial of $K[X,Y]$ and let α, β, λ be three elements of K, with $\alpha\beta \neq 0$, such that the we have

$$(*) \qquad Q(X+\beta, \alpha Y) = \lambda Q(X,Y),$$

then Q belongs to $K[Y]$.

[Consider, first, the case where Q belongs to $K[X]$ and satisfies

$$Q(X+\beta) = \lambda Q(X) \quad \text{with} \quad \beta \neq 0,$$

and then demonstrate that the polynomial Q is constant (otherwise, it would have an infinity of zeros). Then, in the general case, write

$$Q(X,Y) = \sum_{i=0}^{d} a_i(X)\, X^i,$$

and show that condition $(*)$ implies the relations

$$a_i(X+\beta)\,\alpha^i = \lambda a_i(X), \quad \text{for} \quad 0 \leq i \leq d.$$

Conclude.]

Determine the polynomials $Q \in C[Y]$ which satisfy the relation

$$Q(\alpha Y) = \lambda\, Q(Y).$$

[First consider the case when Q is irreducible and satisfies such a relation with $\alpha \neq 1$. Prove that Q is then equal to $Q(Y) = a\,Y$ for a certain element a of K. (After the theorem of d' Alembert — which is proved in the next chapter — the polynomial Q is then linear : $Q(Y) = aY + b$, with $a \neq 0$). From the relation $a\alpha Y + b = \lambda\,(aY+b)$, deduce that $\lambda = \alpha$ (use the fact that $a \neq 0$), then prove that $b = 0$ (use the fact that $\lambda = \alpha \neq 1$). Conclude.]

40. Let K be a field and let $F_1, \ldots, F_r$ be polynomials of $K[X,Y]$, of degree at most L with respect to X and at most M with respect to Y. Let (ξ_i, η_i), for all the indices $1 \leq i \leq N$, be common zeros in K^2 of all polynomials $F_1, \ldots, F_r$, such that the points $\xi_1, \ldots, \xi_N$ are two by two distinct. Prove the inequality $N \leq 2LM$.

[Use the technique of the U-resultant of Kronecker. Introduce the $2r$ new variables $U_1, \ldots, U_r, V_1, \ldots, V_r$ and put

$$G = \sum_{j=1}^{r} U_j\, F_j(X,Y), \qquad H = \sum_{j=1}^{r} V_j\, F_j(X,Y).$$

Let $R \in K[U_1, \ldots, U_r, V_1, \ldots, V_r, X]$ be the resultant with respect to Y of the polynomials F and G. Demonstrate the following facts :

(i) $R \neq 0$,

(ii) $\deg_X R \leq 2LM$,

(iii) $R(U_1, \ldots, U_r, V_1, \ldots, V_r, \xi_i) = 0$ for $i = 1, \ldots, N$.

Then conclude.]

41. From the two exercises above deduce the following result, due also to D. Masser :

Let K be an infinite field, Q a polynomial of $K[X,Y]$, and α, β, γ three nonzero elements of K. Let U_1, U_2, V_1, V_2 be positive integers. Set $U = U_1 + U_2$, and $V = V_1 + V_2$, and suppose that

(a) for $i = 1$ and $i = 2$, the points

$$u + v\beta, \quad (-U_1 \leq u \leq U_1, -V_1 \leq v \leq V_1)$$

are two by two distinct and $(2U_1 + 1)(2V_1 + 1) > L$;

(b) $\operatorname{Card}\{\alpha^u \gamma^v \,;\, -U_1 \leq u \leq U_1, -V_1 \leq v \leq V_1\} > M$;

(c) $\operatorname{Card}\{u + v\beta \,;\, -U_2 \leq u \leq U_2, -V_2 \leq v \leq V_2\} > 2LM$.

Then at least one of the numbers

$$P(u + v\beta, \alpha^u \gamma^v), \quad (-U \leq u \leq U, -V \leq v \leq V)$$

is nonzero.

Moreover, demonstrate that this result is no more true if any of the following conditions does not occur :

(i) $\operatorname{Card}\{(u + v\beta, \alpha^u \gamma^v) \,;\, -U \leq u \leq U,\ -V \leq v \leq V\} \geq (L+1)(M+1)$,

(ii) $\operatorname{Card}\{u + v\beta \,;\, -U \leq u \leq U,\ V \leq v \leq V\} \geq L + 1$,

(iii) $\operatorname{Card}\{\alpha^u \gamma^v \,;\, -U \leq u \leq U,\ -V \leq v \leq V\} \geq M + 1$.

42. Let $f \in \mathbb{R}[X_1, \ldots, X_k]$ be a nonzero polynomial with real coefficients. For any positive integer n, we put

$$\begin{aligned} N_f(x) &= N(x) \\ &= \operatorname{Card}\big\{\mathbf{a} = (a_1, \ldots, a_k) \in \mathbb{Z}^k \,;\, f(\mathbf{a}) = 0 \text{ and } \max\{|a_i| \le x\}\big\}. \end{aligned}$$

1°) Demonstrate that there exists a constant C (which depends on the polynomial f) such that

$$N(x) \le Cx^{k-1}.$$

[Write

$$f = g_m \, {X_1}^m + g_{m-1} \, {X_1}^{m-1} + \cdots + g_0 \,,$$

where $g_i \in \mathbb{R}[X_2, \ldots, X_k]$ for $0 \le i \le m$. Then $g = g_0^2 + \cdots + g_m^2$ is nonzero and

$$N_f(x) \le m(2x+1)^{k-1} + (2x+1)N_g(x) \,.]$$

2°) Show that this estimate is essentially the best possible.

[Examples : $f = X_1 \cdots X_k$, $f = X_1 + \cdots + X_k$.]

3°) How can this result be generalized if we, now suppose f with complex coefficients and $\mathbf{a} \in \left(\mathbb{Z}[\sqrt{-1}]\right)^n$?

43. Let K be a field and let $d_1, \ldots, d_n$ be strictly positive rational integers. Let E be the set of the polynomials $P \in K[X_1, \ldots, X_n]$ whose partial degree with respect to the variable X_i is smaller than d_i for each of the indices $i = 1, 2, \ldots, n$. Show that the application $f : E \longrightarrow K[X]$ defined by

$$f\big(P(X_1, \ldots, X_n)\big) = P(X, X^{d_1}, X^{d_1 d_2}, \ldots, X^{d_1 d_2 \cdots d_{n-1}})$$

is one-to-one. (This result, due to Kronecker, enables us to replace several questions on polynomials in several indeterminates by the case of polynomials in only one variable. Unfortunately, in practice, it has the important drawback of leading to an explosion of the degrees.)

44. Prove that, for any factorial ring A, there are infinitely many irreducible polynomials in $A[X]$.

[Hint : Suppose that F_1, F_2, …, F_n were all these irreducible factors, and consider the polynomial $F_1\,F_2\cdots F_n+1$. Compare with Euclid's argument for the case of prime numbers.]

45. Let A be an integral domain, $F \in A[X_1, X_2, \dots, X_n]$ be a polynomial, and let a_1, a_2, …, a_n be points in A such that $F(a_1, a_2, \dots, a_n) = 0$. Show that there exist polynomials G_1, G_2, …, G_n in $A[X_1, X_2, \dots, X_n]$ such that

$$F = \sum_{i=1}^{n} (X_i - a_i)\, G_i\,.$$

[First, prove that every polynomial F of $A[X_1, X_2, \dots, X_n]$ can be developed as

$$F = \sum \lambda_{\mathbf{i}}\, (X - a_1)^{i_1}\, (X - a_2)^{i_2} \cdots (X - a_n)^{i_n}\,, \quad \lambda_{\mathbf{i}} \in A,$$

where $\mathbf{i} = (i_1, i_2, \dots, i_n) \in \mathbb{N}^n$.]

Are the G_is unique or not ?

46. Let E be any finite set contained in A^n, where A is an integral domain. Construct a nonzero polynomial $F \in A[X_1, X_2, \dots, X_n]$ such that E is contained in the set of the zeroes of F, which means that

$$\{\, (x_1, x_2, \dots, x_n) \in A^n\,;\; F(x_1, x_2, \dots, x_n) = 0 \,\} \supset E\,.$$

[If $Z(F)$ denotes the set of the zeroes of a polynomial, first remark that $Z(F\,G) = Z(F) \cup Z(G)$. Deduce that we may suppose that E contains only one point $(a_1, a_2, \dots, a_n)$. Then consider the product polynomial $(X_1 - a_1)\,(X_2 - a_2) \cdots (X_n - a_n)$.]

47. Let K be a field. Let P_1, P_2, …, P_r be r points belonging to K^n and let λ_1, λ_2, …, λ_n be r elements of K. Demonstrate that there exists a polynomial F in $K[X_1, X_2, \dots, X_n]$ such that $F(P_i) = \lambda_i$ for $i = 1$, 2, …, r.

[First, for any integer $j = 1$, 2, …, r, construct a polynomial F_j such that $F_j(P_i) = \delta_{i\,j}$.]

48. Let K be a field and I be an ideal of the ring $K[X_1, X_2, \ldots, X_n]$. And let P_1, P_r, ..., P_r be r points of K^n such that $F(P_i) = 0$ for every polynomial F in I and all $i = 1, 2, \ldots, r$. Prove the inequality

$$r \leq \dim_K \left(K[X_1, X_2, \ldots, X_n]/I\right).$$

[Prove that the polynomials F_j considered in the preceding exercise, such that $F_j(P_i) = \delta_{ij}$, are linearly independent modulo I.]

49. Let $F(X) = a_0 + a_1 X + a_2 X^2 + \cdots$ be a power series over some field K of characteristic zero. Let N be a positive integer and x_1, x_2, ..., x_N be N points (not necessarily distinct) of K and put

$$R(X) = \prod_{i=1}^{N} (X - x_i).$$

Let m and n be nonnegative integers, with $m + n = N - 1$. Demonstrate that there exist two polynomials P and Q in the ring $K[X]$ which satisfy $\deg P \leq m$, $\deg Q \leq n$ and

$$Q\,F - P \equiv 0 \pmod{R}.$$

50. Let A be a unitary ring and let n be a positive integer that is invertible in A. Let P be a monic polynomial, $P \in A[X]$. Demonstrate that, for every positive integer m, there exists a monic polynomial Q in $A[X]$, of degree m, such that $P - Q^n = 0$ or $\deg(P - Q^n) < mn - m$.

[Put $V = X^m + a_1 X^{m-1} + a_2 X^{m-2} \cdots$, where $a_1, a_2, \ldots$ are unknows. Then, determine these unknows one after another, looking at the terms of degree $mn-1$, $mn-2$, ... of the polynomial $P - Q^n$.]

51. Let C be the following $n \times n$ matrix

$$C = \begin{pmatrix} 0 & 1 & 0 & \ldots & 0 & 0 \\ 0 & 0 & 1 & \ldots & 0 & 0 \\ \ldots & \ldots & \ldots & \ldots & \ldots & \ldots \\ 0 & 0 & 0 & \ldots & 0 & 1 \\ t & 0 & 0 & \ldots & 0 & 0 \end{pmatrix}.$$

1°) Verify that the minimal polynomial of C is $X^n - t$.

2°) Let $a_0, a_1, \ldots, a_{k-1}$ be k elements of some field K in which the polynomial $X^k - t$ splits completely into

$$X^k - t = \prod_{i=1}^{k} (X - \alpha_i).$$

Put

$$A = \begin{pmatrix} a_0 & a_1 & a_2 & \ldots & a_{k-2} & a_{k-1} \\ ta_{k-1} & a_0 & a_1 & \ldots & a_{k-3} & a_{k-2} \\ ta_{k-2} & ta_{k-1} & a_0 & \ldots & a_{k-4} & a_{k-3} \\ \ldots & \ldots & \ldots & \ldots & \ldots & \ldots \\ ta_1 & ta_2 & ta_3 & \ldots & ta_{k-1} & a_0 \end{pmatrix}.$$

Prove the formula

$$\det A = \prod_{i=1}^{k} \left(a_{k-1}\, \alpha_i^{k-1} t^{k-1} + a_{k-2}\, \alpha_i^{k-2} t^{k-2} + \cdots + a_1\, \alpha_i t + a_0\right).$$

3°) Let $P(X)$ be a polynomial with coefficients in some field that contains a primitive kth root of unity ζ (that is, $\zeta^k = 1$ and $\zeta^m \neq 1$ for $1 \leq m < k$). Let P_k be the polynomial whose roots are the kth powers of the roots of P. Demonstrate the formula

$$P_k(X) = \prod_{i=1}^{k} P(\zeta\, X^{1/k}).$$

If $P = \sum_{i=1}^{k} X^i\, A_i(X^k)$, prove that P_k is also equal to the determinant of the matrix

$$\begin{pmatrix} A_0(X) & A_1(X) & A_2(X) & \ldots & A_{k-2}(X) & A_{k-1}(X) \\ X\, A_{k-1}(X) & A_0(X) & A_1(X) & \ldots & A_{k-3}(X) & A_{k-2}(X) \\ X\, A_{k-2}(X) & X\, A_{k-1}(X) & A_0(X) & \ldots & A_{k-4}(X) & A_{k-3}(X) \\ \ldots & \ldots & \ldots & \ldots & \ldots & \ldots \\ X\, A_1(X) & X\, A_2(X) & X\, A_3(X) & \ldots & X\, A_{k-1}(X) & A_0(X) \end{pmatrix}.$$

4°) † Prove the relation

$$P_k(1) = \pm \operatorname{Res}\Big(P(X),\, X^k - 1\Big).$$

† Ph. Glesser, private communication, September 1991.

Deduce the upper bound

$$\mathrm{M}(P) \leq |P|, \quad \text{where } |P| = \max\{|P(z)|\,;\ |z| = 1\}.$$

(The measure $\mathrm{M}(P)$ of P is defined in Section 4.4.) Using Sylvester's matrix for the resultant and Hadamard's inequality, give a new proof of Landau's inequality (see § 4.3.3)

$$\mathrm{M}(P) \leq ||P||,$$

where

$$||P|| = \left(\sum |a_j|^2\right)^{1/2}, \quad \text{if } P = \sum a_j X^j .$$

Deduce also Jensen's formula

$$\mathrm{Log}\, M(P) = \int_0^1 \mathrm{Log}\left|P(e^{2i\pi t})\right| dt .$$

52. Let A be a domain of integrity. For $l = 0, 1, \ldots,$ define the linear operator $E^{(l)}$ on $A[X]$ by the formula

$$E^{(l)}(X^t) = \binom{t}{l} X^{t-l}, \quad \text{for } t = 0, 1, \ldots$$

This operator is sometimes called the *hyperderivative of order* l.

1°) If D is the usual differential operator, prove that $D^l = l\,!\,E^{(l)}$.

2°) Prove the analogue of Leibnitz formula

$$E^{(l)}\left(f_1(X) \cdots f_t(X)\right) = \sum_{i_1 + \cdots + i_t = l} E^{(i_1)}(f_1) \cdots E^{(i_t)}(f_t) .$$

Deduce that, for any c in A, we have

$$E^{(l)}(X - c)^t = \binom{t}{l}(X - c)^{t-l} .$$

3°) Demonstrate that $(X - a)^m$ divides $f(X)$ if, and only if,

$$E^{(l)} f(a) = 0 \quad \text{for } l = 0, 1, \ldots, m - 1 .$$

4°) Using $E^{(l)}$ instead of D^l, generalize the Hermite interpolation polynomial [see Exercise 6 of this chapter] to fields of any characteristic. In the same way, generalize Taylor's formula to any commutative ring.

53. A generalization of van der Monde's formula

1°) Suppose that $C = (c_{i\,j})_{1\leq i,j\leq n}$ is a matrix with elements in some commutative ring A such that there exists an integer $m \geq n$ and elements $a_{i\,s}$, $b_{j\,s}$ in A, $1 \leq i, j \leq n$ and $1 \leq s \leq m$ satisfying the relations

$$c_{i\,j} = \sum_{s=1}^{m} a_{i\,s}\, b_{j\,s}\,, \quad 1 \leq i, j \leq n.$$

If $I = \{i_1, \ldots, i_n\}$ is any subset of cardinality n of the interval $\{1, 2, \ldots, m\}$ we put

$$A_I = \sum \pm a_{1\,i_1} \cdots a_{n\,i_n} \quad \text{and} \quad B_I = \sum \pm b_{1\,i_1} \cdots b_{n\,i_n}\,,$$

where the signs are given by the signatures of the permutations (as for determinants). Then, prove the formula

$$\det(C) = \sum_I A_I\, B_I\,,$$

where the Is run over the subsets of cardinality n of the interval $\{1, 2, \ldots, m\}$.

2°) Let f be monic univariate polynomial of positive degree n, with coefficients in A. For any nonnegative integer i, let s_i be the sum of the ith powers of the roots of f. For any integer ν, $1 \leq \nu \leq n$, set

$$D_\nu = \det(\sigma_{i\,j})\,, \quad \text{where } \sigma_{i\,j} = s_{i+j-2},\ 1 \leq i, j \leq n.$$

If z_1, …, z_n are n indeterminates, put also

$$V(z_1, \ldots, z_n) = \prod_{1\leq i<j\leq n} (z_i - z_j)\,.$$

Using the previous question, demonstrate the formula

$$D_\nu = \sum V^2(\alpha_{i_1}, \ldots, \alpha_{i_\nu})\,,$$

where the sum is extended to all subsets $\{i_1, \ldots, i_\nu\}$ of cardinality ν of the set $\{1, 2, \ldots, n\}$. Deduce the fact that f has exactly r distinct roots if, and only if $D_n = D_{n-1} = \cdots = D_{r+1} = 0$ but $D_r \neq 0$.

Chapter 4

POLYNOMIALS WITH COMPLEX COEFFICIENTS

The main topic of this chapter is the study of the zeros of polynomials with complex coefficients. This study leads to inequalities about the size of factors of polynomials. These inequalities play an important rôle in the last two chapters.

Until Section 4.2 we consider only unitary polynomials.

1. The theorem of d'Alembert

1. Statement

Let us recall that a polynomial is said to be *nonconstant* when its degree is strictly positive. With this definition, the theorem of d'Alembert-Gauss is stated as follows :

THEOREM 4.1. — *Any nonconstant polynomial with complex coefficients has at least one root in the field* $\mathbb{C}$.

By induction on the degree, we can, obviously deduce that any polynomial of positive degree n over $\mathbb{C}$ admits exactly n complex roots* (each of them is counted as often as its order of multiplicity). We say that the field $\mathbb{C}$ is *algebraically closed.*

2. Analytic properties

The only "analytic" properties that will be considered are the following :

(i) Any polynomial of an odd degree with real coefficients admits at least one real root.

(ii) For any positive real number y, there exists a real number x such that $x^2 = y$.

---

* Observe that $f(c) = 0$ implies that $x - c$ divides $f(x)$.

First, note the following lemma.

LEMMA 4.1. — *Let a, b, and c be three elements of a field K, where $a \neq 0$. We suppose that there exists $t \in K$ such that $b^2 - 4ac = t^2$. Then the equation $ax^2 + bx + c = 0$ admits at least one root in K; more precisely, the solutions are*

$$x = \frac{-b \pm t}{2a}.$$

Proof

The demonstration is well known. The equation $ax^2 + bx + c = 0$, where $a, b, c \in K$, $a \neq 0$, can be rewritten as

$$\left(x + \frac{b}{2a}\right)^2 = \frac{b^2 - 4ac}{4a^2}.$$

Hence the solutions

$$x = \frac{-b \pm t}{2a},$$

which are distinct if, and only if, t is nonzero. ☐

Let us show now that property (ii) implies property (ii′) :

(ii′) For any complex number z there is $w \in \mathbb{C}$ such that $w^2 = z$.

Proof

Let $z = a + ib$ be a complex number, a and b real. Let us find w given by $w = u + iv$, where u and v are real. Equation $w^2 = z$ is equivalent to the two conditions $u^2 - v^2 = a$ and $2uv = b$ (hence $-u^2v^2 = -b^2/4$). The numbers u^2 and $-v^2$ are the solutions of the equation $x^2 - ax - b^2/4 = 0$ whose discriminant is $a^2 + b^2$ and hence positive.

Lemma 4.1 and property (ii) show that this last equation admits the solutions

$$u^2 = \frac{a + \sqrt{a^2 + b^2}}{2} \quad \text{and} \quad -v^2 = \frac{a - \sqrt{a^2 + b^2}}{2}.$$

Two new applications of property (ii) give real numbers u and v, which satisfy $(u + iv)^2 = z$. Hence property (ii′). ☐

By applying the lemma again, as well as (ii′), we get the following result.

PROPOSITION 4.1. — *Any second degree equation in the field* $\mathbb{C}$ *has two complex roots (distinct or not).*

3. Demonstration of the theorem

The method is mainly due to Lagrange. We follow the presentation given by Samuel †. Let P be a nonconstant polynomial with complex coefficients. By multiplying P by its complex conjugate, we can suppose that P has real coefficients. [If $(P\bar{P})(a) = 0$, then at least one of the complex numbers a or $\bar{a}$ is a root of P]. Thus, consider a nonconstant polynomial P, with real coefficients. Suppose that the degree of P is $d = 2^n q$, with q odd.

We argue by induction on the integer n. If n is zero, then property (i) implies that P has a real root. Suppose that $n \geq 1$. Let $x_1, \ldots, x_d$ be the roots of P in a suitable extension of $\mathbb{C}$.

For any real number c, we write

$$y_{hk}(c) = y_{hk} = x_h + x_k + cx_h x_k,$$

for $1 \leq h \leq k \leq d$. The coefficients of the polynomial

$$Q(X) = \prod_{h<k} (X - y_{hk})$$

are symmetrical polynomial functions with respect to the x_i. Since the coefficients of P are real, the coefficients of Q are real too.

We have $\deg(Q) = d(d-1)/2 = 2^{n-1}q(d-1)$, with $q(d-1)$ odd. Thus we can apply the induction hypothesis to polynomial Q : there exists a root of Q in the field $\mathbb{C}$, hence a pair of indices (h, k) such that $y_{hk}(c)$ belongs also to $\mathbb{C}$.

Using $d(d-1)/2 + 1$ different values of the parameter c, we see that there is a pair of indices (r, s) and two distinct real numbers c and c', such that the numbers $y_{rs}(c)$ and $y_{rs}(c')$ are in $\mathbb{C}$. Hence, the sum $x_r + x_s$ and the product $x_r x_s$ belong to $\mathbb{C}$ too. Hence, x_r and x_s are solutions of a second-degree equation with complex coefficients, and by Proposition 4.1 they belong to $\mathbb{C}$. Hence, the conclusion. □

† P. Samuel, *Théorie des Nombres Algébriques*, Hermann, Paris, 1968, Appendix, Chapter 2.

Remark

We have, indeed, proved that, if $\mathbb{R}$ is an ordered field that satisfies (i) and (ii), then the field $\mathbb{R}[\sqrt{-1}]$ is algebraically closed.

4. Irreducible real polynomials

Theorem 4.1 has the following consequence, whose proof will be seen in an exercise.

PROPOSITION 4.2. — *The only irreducible polynomials of* $\mathbb{R}[X]$ *are first-degree polynomials and those of the second degree whose discriminant is negative.*

2. Estimates of the roots

Most of the results in this section are due to Cauchy. The presentation given here, is based on the book *Calcul Infinitésimal* by J. Dieudonné‡.

1. Principle of the demonstration

PROPOSITION 4.3. — *Let* $P(X) = X^d + a_{d-1}X^{d-1} + \cdots + a_0$ *be a polynomial with complex coefficients whose roots are* $z_1, \ldots, z_d$. *Let us write*

$$r_0 = \max\{|z_1|, |z_2|, \ldots, |z_d|\}.$$

Then, if a real number $r > 0$ *satisfies the condition*

$$r^d \geq |a_{d-1}|r^{d-1} + \cdots + |a_0|,$$

we have the inequality $r_0 \leq r$.

Proof

Suppose $r_0 > 0$, otherwise there is nothing to demonstrate. According to Lemma 4.2 below, the polynomial

$$X^d - |a_{d-1}|X^{d-1} - \cdots - |a_0|$$

has only one positive real positive root, call it r_1. It is obvious that r_0 is at most equal to r_1.

‡ Hermann, Paris, 1968, (Prob. 1, Ch. 2).

Moreover, for $j = 1, \ldots, d$, we have

$$|z_j^d| = |a_{d-1}z_j^{d-1} + \cdots + a_0| \leq |a_{d-1}||z_j^{d-1}| + \cdots + |a_0|,$$

and, in particular

$$r_0^d \leq |a_{d-1}|r_0^{d-1} + \cdots + |a_0|.$$

We can deduce the inequalities $r_0 \leq r_1 \leq r$. $\square$

2. An analytic lemma

The result used above is the following.

LEMMA 4.2. — *Any polynomial with real coefficients that is given by a formula as*

$$c_d X^d + c_{d-1}X^{d-1} + \cdots + c_j X^j - c_{j-1}X^{j-1} - \cdots - c_0, \quad \textit{with} \quad d \geq 1,$$

where $c_i \geq 0$ *for* $i = 0, \ldots, d$ *and* $c_d + \ldots + c_j > 0$, $c_{j-1} + \ldots + c_0 > 0$, *admits a real positive root and only one.*

Proof

The function

$$f(x) = c_d x^{d-j} + c_{d-1}x^{d-1-j} + \ldots + c_j - c_{j-1}x^{-1} - \cdots - c_0 x^{-j}$$

is continuous on the real interval $]0, \infty[$, strictly increasing (as a sum of strictly increasing functions) and tends to $-\infty$ (and respectively, $+\infty$) when x tends to zero (respectively, to $+\infty$); hence, it admits one and only one positive zero. $\square$

Remark

This lemma is a particular case of the rule of Descartes (for the general case of Descartes' rule, see Chapter 5, § 4.3).

3. Bounds for the roots

Proposition 4.3 has the following corollary.

THEOREM 4.2. — *Let* $P(X) = X^d + a_{d-1}X^{d-1} + \cdots + a_0$ *be a polynomial with complex coefficients and whose roots are* $z_1, \ldots, z_d$.

Let us write

$$r_0 = \max\{|z_1|, |z_2|, \ldots, |z_d|\}.$$

Then r_0 *satisfies the following inequalities :*

(i) $$r_0 \leq \max\left\{1, \sum_{k=0}^{d-1} |a_k|\right\},$$

(ii) $$r_0 \leq 1 + \max_{0 \leq k < d} \{|a_k|\},$$

(iii) *if* $\lambda_1, \ldots, \lambda_d$ *are positive real numbers satisfying the condition* $\lambda_1^{-1} + \cdots + \lambda_d^{-1} = 1$ *then we have*

$$r_0 \leq \max_{1 \leq k \leq d} \left\{\left(\lambda_k |a_{d-k}|\right)^{1/k}\right\},$$

(iv) $$r_0 \leq \max_{1 \leq k \leq d} \left\{\left(d|a_{d-k}|\right)^{1/k}\right\},$$

(v) *if all the coefficients* a_k *are nonzero, we get*

$$r_0 \leq \max\left\{2|a_{d-1}|, 2\left|\frac{a_{d-2}}{a_{d-1}}\right|, \ldots, 2\left|\frac{a_1}{a_2}\right|, \left|\frac{a_0}{a_1}\right|\right\},$$

(vi) $$r_0 \leq |1 - a_{d-1}| + |a_{d-2} - a_{d-1}| + \cdots + |a_0 - a_1| + |a_0|,$$

(vii) *if all the* a_k *are real and positive, then*

$$r_0 \leq \max\left\{a_{d-1}, \frac{a_{d-2}}{a_{d-1}}, \ldots, \frac{a_1}{a_2}, \frac{a_0}{a_1}\right\},$$

(viii) *finally*

$$r_0 \leq \max_{1\leq k\leq d} \left\{ \left(\frac{a_{d-k}}{\binom{d}{k}} \right)^{1/k} (2^{1/d} - 1)^{-1} \right\}.$$

Proof

(i) If $|a_{d-1}|+\ldots+|a_0| \leq 1$, choose $r = 1$ in Proposition 4.3, otherwise, take $r = |a_{d-1}| + \ldots + |a_0|$.

(ii) Let A be the maximum of the $|a_k|$ for $0 \leq k < d$. We can take $r = A + 1$, indeed

$$r^d = A(r^{d-1} + \cdots + 1) + 1 > |a_{d-1}|r^{d-1} + \cdots + |a_0|.$$

(iii) Let us call ρ the right-hand side of inequality (iii). Then

$$|a_{d-k}| \leq \lambda_k^{-1}\rho^k, \quad \text{for } 1 \leq k \leq d.$$

Hence the inequality

$$\sum_{k=1}^{d} |a_{d-k}|\rho^{d-k} \leq \sum_{k=1}^{d} \rho^d \lambda_k^{-1} = \rho^d.$$

Therefore, we can take $r = \rho$.

(iv) Take each of the λ_k equal to d in inequality (iii). (Remark : we can even take all the λs equal to λ = number of nonzero a_is. Prove this remark as an exercise.)

(v) Again, let us call ρ the right-hand side. We successively demonstrate the inequalities

$$|a_{d-1}| \leq \rho/2,\ |a_{d-2}| \leq (\rho/2)^2, \ldots,\ |a_1| \leq (\rho/2)^{d-1},\ |a_0| \leq 2(\rho/2)^d.$$

Hence we obtain the inequality

$$\sum_{k=1}^{d} |a_{d-k}|\,\rho^{d-k} \leq \rho^d$$

and the conclusion, $\rho \geq r$.

(vi) Apply (i) to the polynomial $(X-1)\,P(X)$.

(vii) Let ρ be the right-hand side of inequality (vii). Through the change of variables $X \mapsto X/\rho$, polynomial P becomes

$$\rho^d X^d + a_{d-1}\rho^{d-1}X^{d-1} + \cdots + a_0 .$$

When applying relation (vi) to this polynomial, we get

$$\frac{r_0}{\rho} \le \left|1 - \frac{a_{d-1}}{\rho}\right| + \left|\frac{a_{d-1}}{\rho} - \frac{a_{d-2}}{\rho^2}\right| + \cdots + \left|\frac{a_0}{\rho^d}\right| .$$

To conclude, just notice that the right-hand side of this last inequality is just equal 1 (indeed, we get $a_0 \le \rho^d$).

(viii) : Apply (iii) with

$$\lambda_k^{-1} = \binom{d}{k}(2^{1/d}-1)^k .$$

Hence the result. ☐

Remark

Considering the polynomial $X^d P(X^{-1})$, called the *reciprocal polynomial* of P, and supposing that the constant term a_0 is not zero, we get lower bounds for $|z_j|$. For example,

$$\min\{|z_1|, |z_2|, \ldots, |z_d|\} \ge \frac{a_0}{1 + \max\limits_{1\le k\le d}\{|a_{d-k}|\}} .$$

3. The measure of a polynomial

1. Definition

Let P be a polynomial with complex coefficients. Let us write this polynomial as

$$P = a_d X^d + a_{d-1}X^{d-1} + \cdots + a_0 = a_d(X-z_1)\cdots(X-z_d),$$

with $a_d \ne 0$.

The *measure* of P is defined by the formula

$$\mathrm{M}(P) = |a_d| \prod_{j=1}^{d} \max\{1, |z_j|\}.$$

This notion has been introduced by K. Mahler. If P and Q are nonzero polynomials, where P is of degree d, notice that

$$\mathrm{M}\big(X^d P(1/X)\big) = \mathrm{M}(P) \quad \text{and} \quad \mathrm{M}(P \cdot Q) = \mathrm{M}(P) \cdot \mathrm{M}(Q),$$

and also that, for any positive integer k, we have

$$\mathrm{M}\big(P(X^k)\big) = \mathrm{M}\big(P(X)\big).$$

It is time to introduce some definitions. Let A be a subring of an integral domain B. An element x of B is said to be *algebraic* over A when it is the root of a nonzero polynomial with coefficients in A; then there exists a nonzero polynomial of $A[X]$ — of minimal degree and whose coefficients are relatively prime — whose value at x is zero; such a polynomial is called a *minimal polynomial* (over A) of the element x.

A complex number that is algebraic over the field $\mathbb{Q}$ is simply called an *algebraic number*. When α is a nonzero algebraic number, and by definition its measure $\mathrm{M}(\alpha)$ is given by the formula $\mathrm{M}(\alpha) = \mathrm{M}(P)$, where P is any minimal polynomial of α over $\mathbb{Z}$. (Observe that a minimal polynomial over $\mathbb{Z}$ is unique up to a multiplication by -1. More generally, a minimal polynomial over a ring A is unique up to a multiplication by an invertible element of A.)

The measure of algebraic numbers satisfies some important properties, let us quickly review them. Let α and α' be two nonzero algebraic numbers, of respective degrees d and d'. The absolute value $|\alpha|$ satisfies

$$\text{(M1)} \qquad \frac{1}{\mathrm{M}(\alpha)} \leq |\alpha| \leq \mathrm{M}(\alpha).$$

The measure of the product $\alpha \cdot \alpha'$ satisfies

$$\text{(M2)} \qquad \mathrm{M}(\alpha \cdot \alpha') \leq M(\alpha)^{d'} \cdot M(\alpha')^{d},$$

and the measure of the sum $\alpha + \alpha'$ satisfies

$$\text{(M3)} \qquad \mathrm{M}(\alpha + \alpha') \leq 2^{dd'}\,\mathrm{M}(\alpha)^{d'} \cdot M(\alpha')^{d}.$$

Let k be a positive integer, then

$$\text{(M4)} \qquad \mathrm{M}\left(\alpha^{1/k}\right) \leq M(\alpha).$$

and

$$\text{(M5)} \qquad \mathrm{M}(\alpha^{-1}) = \mathrm{M}(\alpha) \quad \text{and} \quad \mathrm{M}(\alpha^k) \leq M(\alpha)^k.$$

Proof

(M1) The proof is almost obvious.

(M2) Let a and a' be the respective leading coefficients of the minimal polynomials of α and α'. Let $\{\alpha_i \; ; \; 1 \leq i \leq d\}$ and $\{\alpha'_j \; ; \; 1 \leq j \leq d'\}$ be the respective sets of the conjugates of α and of α', then $\alpha \cdot \alpha'$ is a root of the polynomial

$$a^{d'} \cdot a'^d \cdot \prod_{i,j} (z - \alpha_i \cdot \alpha'_j),$$

moreover, this polynomial has integer coefficients. Which implies the estimates

$$\begin{aligned} \mathrm{M}(\alpha.\alpha') &< a'^d a^{d'} \cdot \prod_{i,j} \max\{1, |\alpha_i \cdot \alpha'_j|\} \\ &\leq a'^d a^{d'} \cdot \Big(\prod_i \max\{1, |\alpha_i|\}\Big)^{d'} \cdot \Big(\prod_j \max\{1, |\alpha'_j|\}\Big)^d \\ &= \mathrm{M}(\alpha)^{d'} \cdot \mathrm{M}(\alpha')^d. \end{aligned}$$

(M3) The proof is the same as above.

(M4) Let P be the minimal polynomial of α over $\mathbb{Z}$. Then $\alpha^{1/k}$ is a root of the polynomial $Q(X) := P(X^k)$, so that

$$\mathrm{M}\left(\alpha^{1/k}\right) \leq \mathrm{M}(Q) = \mathrm{M}(P) = \mathrm{M}(\alpha).$$

(M5) The proofs are easy, and we leave them to the reader. $\square$

2. An algebraic lemma

For a polynomial P with complex coefficients given by

$$P = a_d X^d + a_{d-1} X^{d-1} + \cdots + a_0 = a_d (X - z_1) \cdots (X - z_d),$$

introduce the following notation

$$||P|| = ||P||_2 = \left(|a_0|^2 + \cdots + |a_d|^2\right)^{1/2}.$$

The following elementary result will allow us to find upper bounds for the measure of a polynomial.

LEMMA 4.2. — *For any polynomial Q with complex coefficients and for any complex number z, we have the relation*

$$||(X - z)Q(X)|| = ||(\bar{z}X - 1)Q(X)||.$$

Proof

Let us suppose that the polynomial Q is written $Q = \sum_{k=0}^{m} c_k X^k$. Then the square of the left-hand side admits the development

$$\sum_{k=0}^{m+1} (c_{k-1} - z c_k)\,(\bar{c}_{k-1} + \bar{z}\bar{c}_k) =$$

$$(1 + ||z||^2)\,||Q||^2 - \sum_{k=0}^{m} (z c_k \bar{c}_{k-1} + \bar{z}\bar{c}_k c_{k-1}),$$

where we have put $c_{-1} = c_{m+1} = 0$. We immediately verify that the square of the right-hand side admits exactly the same development. ☐

3. An upper bound for $\mathrm{M}(P)$

Theorem 4.2 leads immediately to upper bounds for the measure of a polynomial, but it is possible to get an estimate, which is often sharper using Theorem 4.3 below.

THEOREM 4.3 (Vicente Gonçalves). — *Let*

$$P = a_d X^d + a_{d-1} X^{d-1} + \cdots + a_0$$

be a nonconstant polynomial with complex coefficients. The measure of P satisfies the inequality

$$\mathrm{M}(P)^2 + |a_0 a_d|^2\, \mathrm{M}(P)^{-2} \leq ||P||^2.$$

Proof

Let $z_1, \ldots, z_k$ be the roots of P which are outside the unit disc. Then the measure of P is $\mathrm{M}(P) = |a_d|\,|z_1 \cdots z_k|$.

Let us consider the polynomial

$$R(X) = a_d \prod_{j=1}^{k} (\bar{z}_j X - 1) \cdot \prod_{j=k+1}^{d} (X - z_j) = b_d X^d + \cdots + b_0 .$$

Applying the relation of the previous lemma k times, we get the relation $||P|| = ||R||$. But, on the other hand, we have

$$||R||^2 \geq |b_d|^2 + |b_0|^2 = \mathrm{M}(P)^2 + |a_d a_0|^2 M(P)^{-2} .$$

Hence, we reach the conclusion. □

COROLLARY (inequality of Landau). — *If the polynomial P is not reduced to a monomial, we have the inequality*

$$\mathrm{M}(P) < ||P|| .$$

Remark

In the inequality of Theorem 4.3, equality holds when the polynomial is equal to $X^n - 1$.

Note : The inequality of the corollary appeared already in an article of Landau*. The theorem given here has been demonstrated by Vicente Gonçalves, and independently by the author†. The demonstration given here is a simplification of the proof of this last article.

4. Other upper bounds

Let z be any complex number; then, by a computation similar to that of lemma 4.3, we have

$$\big|\big| (X + z) P(X) \big|\big|^2 = (1 + |z|^2)\, ||P||^2 + 2\,\mathrm{Re}(za) ,$$

---

* Sur quelques théorèmes de M. Petroviċ relatifs aux zéros des fonctions analytiques; *Bull. Soc. Math. France*, 33, 1905, p. 251–261.

† An inequality about factors of polynomials; *Math. of Computation*, 28, 1974, p. 1153–1157.

where $\operatorname{Re}(x)$ designates the real part of the complex number x, and a is given by the formula

$$a = \sum_{k=1}^{d} a_k \bar{a}_{k-1}.$$

By choosing $z = -\bar{a}||P||^{-2}$, we get the relation

$$\left|\left|(X - \bar{a}||P||^{-2})P(X)\right|\right|^2 = ||P||^2 - |a|^2||P||^{-2}.$$

Also, for any z, we have

$$\mathrm{M}\big((X+z)\cdot P\big) \geq \mathrm{M}(P).$$

Applying the previous corollary, we get the following estimate.

Proposition 4.4. — *Let $P = a_d X^d + a_{d-1}X_{d-1} + \cdots + a_0$ be a nonconstant polynomial with complex coefficients. The measure of P satisfies the inequality*‡

$$\mathrm{M}(P)^2 \leq ||P||^2 - |a_d\bar{a}_{d-1} + a_{d-1}\bar{a}_{d-2} + \cdots + a_1\bar{a}_0|^2||P||^{-2}.$$

Remark

Let j be a positive integer. Then, the same demonstration, applied to the polynomial $(X^j + z)P(X)$, gives the upper bound

$$\mathrm{M}(P)^2 < ||P||^2 - \max_{1\leq j\leq d}\left\{|\bar{a}_0 a_j + \bar{a}_1 a_{j+1} + \cdots \bar{a}_{d-j}a_d|^2\right\} \times ||P||^{-2},$$

when the polynomial P is not reduced to a monomial.

5. *Analytic results*

Jensen's formula implies that the measure of a polynomial is also given by the expression

$$\mathrm{M}(P) = \exp\left\{\frac{1}{2\pi}\int_0^{2\pi} \operatorname{Log}\left|P(e^{i\theta})\right| d\theta\right\}.$$

‡ This inequality has been demonstrated in the author's article — Entiers algébriques dont les conjugués sont proches du cercle unité, *Seminaire Delange-Pisot-Poitou*, 1977/78, exposé 39 — but with a more complicated method.

A general theorem of Szegö¶ leads to the following result.

PROPOSITION 4.5. — *Let P be a nonconstant polynomial with complex coefficients. Then, we have*

$$\mathrm{M}(P) = \mathit{inf}\,\{||PQ||\,;\,Q \in \mathbb{C}[X],\ \mathit{monic}\}.$$

Moreover, for each integer n, it is possible to compute the quantity

$$I_n(P) := \mathit{inf}\,\{||PQ||\,;\,Q \in \mathbb{C}[X],\ \mathit{monic},\ \deg(Q) = n\}.$$

Proof

In other words, we have

$$\mathrm{M}(P) \leq I_n(P) \quad \text{for all } n \geq 0,$$

which is a direct consequence of Theorem 4.3 above. And we also have $I_n(P) \to M(P)$ when $n \to \infty$, which is the difficult part of the theorem of Szegö. □

The inequality $\mathrm{M}(P) \leq I_0(P)$ is the corollary of Theorem 4.3, whereas the inequality $\mathrm{M}(P) \leq I_1(P)$ corresponds to proposition 4.4.

Also, mention the theorem of Rouché (see, for example, the book of Marden§ of that of Henrici†. which is often useful to locate the roots of a polynomial.

PROPOSITION 4.6 (Theorem of Rouché). — *When a nonconstant polynomial*

$$c_d z^d + c_{d-1} z^{d-1} + \cdots + c_0,$$

with complex coefficients satisfies the condition

$$|c_j| > |c_0| + \cdots + |c_{j-1}| + |c_{j+1}| + \cdots + |c_d|$$

for some index j, $0 \leq j \leq d$, this polynomial has exactly j roots with modulus < 1 and no roots with modulus equal to 1.

¶ See, for example, the book of H. Dym and H.P. McKean, *Fourier Series and Integrals*, Academic Press, London, 1972.

§ M. Marden, *The geometry of the zeros of a polynomial in a complex variable*, American Mathematical Society, New York, 1949.

† *Computational complex analysis*, Addison-Wesley, New York, 1976.

6. A method for the computation of $\mathrm{M}(P)$

This method is adapted from the classical method of Dandelin-Graeffe, that is used in Numerical Analysis for the computation of the biggest modulus of the roots of a polynomial with complex coefficients.

To a polynomial P satisfying the formula

$$P(X) = a_d \prod_{j=1}^{d} (X - z_j),$$

we associate a sequence of polynomials P_m such that

$$P_m(X) = \pm\, {a_d}^{2^m} \cdot \prod_{j=1}^{d} (X - {z_j}^{2^m}), \quad m \geq 0,$$

and which can be very easily computed as follows.

Write $P_0 = P$, and, if the polynomial P_m can be written as

$$P_m(X) = F_m(X^2) + X G_m(X^2),$$

then, the polynomial P_{m+1} is given by the formula

$$P_{m+1}(X) = {F_m}^2(X) - X {G_m}^2(X).$$

It is obvious that these polynomials satisfy $\mathrm{M}(P_m) = \mathrm{M}(P)^{2^m}$. Moreover, on one hand, we have the upper bound $\mathrm{M}(P_m) \leq ||P_m||$, and, on the other hand, the inequality $||P_m|| \leq 2^d\, \mathrm{M}(P_m)$. Hence we have the following result.

PROPOSITION 4.7. — *Let P be a polynomial of degree $d \geq 1$ with complex coefficients and let P_m, $m \geq 0$, be the sequence of polynomials associated with it in the Graeffe's method. The measure of P satisfies the inequalities*

$$2^{-d2^{-m}} \cdot ||P_m||^{2^{-m}} \leq \mathrm{M}(P) \leq ||P_m||^{2^{-m}}.$$

Thus

$$||P_m||^{2^{-m}} \longrightarrow \mathrm{M}(P)$$

when $m \to \infty$. *

---

* This method to compute the measure of a polynomial has been discovered independently and almost simultaneously by several people (including D. Boyd, M. Langevin, C. Stewart, and M. Mignotte).

The presentation given in this section, as well as the example that follows are inspired the article of L. Cerlienco, M. Mignotte and F. Piras †.

7. An example of the evaluation of $\mathrm{M}(P)$

Let us consider, for example, a six-degree polynomial, namely the polynomial

$$P(X) = X^6 + X^5 + 6X^4 - 5X^3 + 3X^2 + 2.$$

Now apply some of the estimates given above. The Corollary of Theorem 4.3 gives the rather crude upper bound

$$\mathrm{M}(P) \leq ||P|| = (76)^{1/2} < 8.718,$$

whereas Proposition 4.4 implies

$$\mathrm{M}(P) \leq \Big(76 - (-15 - 30 + 6 + 1)^2/76\Big)^{1/2} < 7.55.$$

By applying the method of Dandelin-Graeffe, starting from P, we, successively, obtain the polynomials

$$P_1(X) = X^6 + 11X^5 + 52X^4 + 15X^3 + 33X^2 + 12X + 4,$$
$$P_2(X) = X^6 - 17X^5 + 2440X^4 + 2951X^3 + 1145X^2 + 120X + 16,$$

and

$$P_3(X) = X^6 + 4591X^5 + 6056224X^4 - 3116689X^3 + 680865X^2 + 22240X + 256.$$

If we apply Proposition 4.7 for $m = 3$, we get

$$2^{-6/8}||P_3||^{1/8} \leq \mathrm{M}(P) \leq ||P_3||^{1/8},$$

so that $4.252 < \mathrm{M}(P) < 7.152$.

† Computing the measure of a polynomial, *J. Symb. Comp.*, 4 (1) 1987, p. 21–34.

The application of Proposition 4.4 to the polynomial P_3 gives the upper bound

$$M(P) < 7.053.$$

In fact, we are much more interested by an upper bound of the measure of P than by a lower bound of this measure. Therefore, we stop the computation at the third step. Nevertheless, we can easily sharpen the lower bound of $\mathrm{M}(P)$: looking at the third coefficient of P_3, we get

$$6056224 \leq \binom{6}{2} \mathrm{M}(P_3),$$

so that the measure of P satisfies

$$\mathrm{M}(P) \geq (6056224/15)^{1/8} > 5.02.$$

It is easy to prove that the polynomial P has no real root of absolute value ≥ 1. Also, the theorem of Rouché shows that P_3 — hence also the polynomial P — has exactly two roots of modulus > 1. Thus, the polynomial P has two complex conjugate roots of modulus > 1 (since P is monic, it is irreducible over the ring $\mathbb{Z}$).

This information leads to a more precise estimate of $\mathrm{M}(P)$. Therefore, let us consider, again, the third coefficient $b := 6056224$ of the polynomial P_3. This coefficient is the sum of all the products of distinct pairs of roots of P_3, hence, it satisfies the inequality

$$M^8 - 8M^4 - 6 \leq b \leq M^8 + 8M^4 + 6,$$

where M designates the measure of the polynomial P, which implies the estimate

$$(b+10)^{1/2} - 4 < M^4 < (b+22)^{1/2} + 4,$$

and finally

$$7.040 < M(P) < 7.047.$$

4. The size of the factors of a polynomial

1. Definitions

If $R = c_d X^d + \cdots + c_0$ is a polynomial with complex coefficients, we define its *length* by the formula

$$\mathrm{L}(R) = \sum_{k=0}^{d} |c_k| .$$

We also define the *height* of R by

$$\mathrm{H}(R) = \max\{|c_0|, |c_1|, \ldots, |c_d|\} .$$

If $z_1, \ldots, z_d$ designate the roots of the polynomial R, counted with their multiplicities, the formula $R = c_d(X - z_1)\cdots(X - z_d)$ shows that

$$\frac{c_{d-j}}{c_d} = (-1)^{d-j} \sum_{i_1 < \ldots < i_j} z_{i_1} \cdots z_{i_j} \quad \text{for } j = 1, 2, \ldots, d .$$

Hence the inequalities

$$|c_j| \le \binom{d}{j} \mathrm{M}(R) , \quad \mathrm{H}(R) \le \binom{d}{[d/2]} \mathrm{M}(R) ,$$

and

$$\mathrm{L}(R) \le 2^d \, \mathrm{M}(R) .$$

2. Upper bounds for the factors, the case of a single variable

The fundamental result is the following.

THEOREM 4. — *Let $P \in \mathbb{Z}[X]$ be a nonconstant polynomial that admits the factorization*

$$P = Q_1 \cdots Q_m Q , \qquad Q_1, \ldots, Q_m, Q \in \mathbb{Z}[X] .$$

Then, we have

$$\prod_{j=1}^{m} \mathrm{L}(Q_j) \le 2^D \, \mathrm{M}(P) , \qquad \text{where } D = \sum_{j=1}^{m} \deg(Q_j) ,$$

and thus

$$\prod_{j=1}^{m} \mathrm{L}(Q_j) \le 2^D ||P|| .$$

Proof

After the previous paragraph, we have

$$\prod_{j=1}^{m} \mathrm{L}(Q_j) \leq 2^D \mathrm{M}(Q_1) \cdots \mathrm{M}(Q_m).$$

Then the inequalities

$$\mathrm{M}(Q_1) \cdots \mathrm{M}(Q_m) \leq \mathrm{M}(P) \leq ||P||$$

lead to the conclusion. □

3. Definition of the measure in the case of several variables

Now, let $P(X_1, \ldots, X_n)$ be a polynomial in n variables, $n \geq 2$. We define recursively the measure of P by the formula

$$\operatorname{Log} \mathrm{M}(P) = \int_0^1 \operatorname{Log}\left\{\mathrm{M}\left(P(e^{2i\pi t}, X_2, \ldots, X_n)\right)\right\} dt.$$

Following the formula of Section 3.5, we find that this expression is also correct for $n = 1$. By induction on n, deduce that the measure has the property of multiplicativity :

$$\mathrm{M}(P \cdot Q) = \mathrm{M}(P) \cdot \mathrm{M}(Q),$$

which had already been noted in the case of one variable. If the polynomial $P(X_1, \ldots, X_n)$ is given by the formula

$$P = \sum a_{j_1 \ldots j_n} X_1^{j_1} \cdots X_n^{j_n},$$

we also write

$$||P|| = \left(\sum |a_{j_1 \ldots j_n}|^2\right)^{1/2}.$$

Let us suppose that we have demonstrated the inequality $\mathrm{M}(R) \leq ||R||$ in the case of polynomials in $n - 1$ variables, then

$$\operatorname{Log} \mathrm{M}(P) \leq \tfrac{1}{2} \int_0^1 \operatorname{Log}\left\{||P(e^{2i\pi t}, X_2, \ldots, X_n)||^2\right\} dt.$$

The inequality of Jensen[‖].

$$\int_0^1 \operatorname{Log}|f(t)|\,dt \le \operatorname{Log}\Big(\int_0^1 |f(t)|\,dt\Big),$$

then gives the inequality

$$\mathrm{M}(P)^2 \le \int_0^1 \big|\big|P(e^{2i\pi t}, X_2, \dots, X_n)\big|\big|^2\,dt.$$

Thanks to the formula of Parseval,

$$\begin{aligned}\int_0^1 \big|\big|P(e^{2i\pi t}, X_2, \dots, X_n)\big|\big|^2\,dt &= \sum_{j_2,\dots,j_n} \int_0^1 \Big|\sum_{j_1} a_{j_1,j_2,\dots,j_n} e^{2i\pi j_1 t}\Big|^2\,dt \\ &= ||P||^2,\end{aligned}$$

we deduce, again, the upper bound $\mathrm{M}(P) \le ||P||$.

4. Bounds for the factors of a polynomial, case of several variables

Let us start with the bounds of coefficients of a polynomial in terms of its measure.

PROPOSITION 4.8. — *Let $P(X_1, \dots, X_n)$ be a polynomial with complex coefficients which can be written*

$$P = \sum a_{j_1 \dots j_n} X_1^{j_1} \cdots X_n^{j_n}.$$

Then, we have the inequalities

$$|a_{j_1 \dots j_n}| \le \binom{d_1}{j_1} \cdots \binom{d_n}{j_n} \mathrm{M}(P),$$

where d_j designates the partial degree of P relative to the variable X_j, for all $j = 1, 2, \dots, n$.

‖ See, for example, G.H. Hardy, J.E. Littlewood, and G. Pólya, *Inequalities*, Cambridge University Press, Cambridge, 1934.

Proof

We have seen in Section 4.1 that these inequalities are true for $n = 1$. Now, let us suppose $n \geq 2$. Then, for any $z_2, \ldots, z_n$, we have

$$\sum_{j_2,\ldots,j_n} a_{j_1\, j_2 \ldots j_n} {z_2}^{j_2} \cdots {z_n}^{j_n} \leq \binom{d_1}{j_1} \mathrm{M}\{||P(X, z_2, \ldots, z_n)||\}.$$

Let us designate by P_{j_1} the polynomial in $X_2, \ldots, X_n$ corresponding to the left-hand side. By integrating the logarithm of each member of the above inequality on the complex domain $\{|z_2| = \cdots = |z_n| = 1\}$, we get the inequality

$$\mathrm{M}(P_{j_1}) \leq \binom{d_1}{j_1} \mathrm{M}(P).$$

Arguing by induction on n, we can write

$$|a_{j_1 \ldots j_n}| \leq \binom{d_2}{j_2} \cdots \binom{d_n}{j_n} \mathrm{M}(P_{j_1}).$$

Hence, we have the result. $\square$

By putting together the results of the two Sections above, we get the following theorem.

THEOREM 4.4 bis. — *Let $P(X_1, \ldots, X_n)$ be a polynomial with complex coefficients whose degree relative to each of the variables X_j is equal to d_j. Then, the length of P satisfies*

$$\mathrm{L}(P) \leq 2^{d_1 + \cdots + d_n} \mathrm{M}(P) \leq 2^{d_1 + \cdots + d_n} ||P||.$$

Moreover, if $P = Q_1 \cdots Q_m$, we have

$$\mathrm{L}(Q_1) \cdots \mathrm{L}(Q_m) \leq 2^{d_1 + \cdots + d_n} \mathrm{M}(P) \leq 2^{d_1 + \cdots + d_n} ||P||.$$

Remark

The first of the inequalities of this type seem to be due to Gel'fond†.

† A.O. Gel'fond .— *Transcendental and Algebraic Numbers*, Dover, New York, 1960, Ch. 3, Sec. 4.

5. *An example in the case of one variable*

The aim of the construction that follows is to show that the inequality

$$\mathrm{L}(Q) \leq 2^{\deg(Q)}||P||,$$

when P and Q are polynomials with integer coefficients and Q is a divisor of P, is almost the best possible. Hence, we have to show that there exist nonzero polynomials P and Q with integer coefficients and where P has "small" coefficients and Q has a "big" length.

Therefore, we choose some polynomial Q with a large length. Namely, we take $Q = (X-1)^n$, so that the length of Q is equal to 2^n. Now, we have to find a polynomial P (nonzero!) with integer coefficients, which is a multiple of Q and has "small" coefficients.

We search P of degree at most d (to be determined later) and of height equal to 1, that is with coefficients 0, or ± 1. The construction of such a polynomial, of degree "not too big" — or the proof of the fact that such a polynomial exists — will use the "pigeon hole principle".

Thus, let n be a fixed positive integer and d an integer that will be chosen later. The fact that the polynomial P is a multiple of Q is equivalent to all the conditions

$$P(1) = P'(1) = \cdots = P^{(n-1)}(1) = 0.$$

If $P = a_0 + \cdots + a_d X^d$, where the coefficients a_i are unknown, this is, also, equivalent to the linear system

$$\sum_{i=0}^{d} \binom{i}{t} a_i = 0, \quad t = 0, 1, \ldots, n-1.$$

Let E be the set of polynomials of degree at most d and with coefficients equal to 0 or 1. The application f of E in the set $\mathbb{N}^{n+1}$ defined by

$$(c_0 + \cdots + c_d X^d) \longmapsto \left(\sum_{i=0}^{d} \binom{i}{t} c_i\right)_{0 \leq t \leq n}$$

sends the set E, which has 2^{d+1} elements, on the set $F = f(E)$ whose number of elements is at most

$$(d+1)^{1+2+\cdots+n} = (d+1)^{n(n+1)/2}.$$

Indeed, for $t = 0, 1, \ldots, n-1$, we have

$$c_i \in \{0,1\} \Longrightarrow \sum_{i=0}^{d} \binom{i}{t} c_i \leq \sum_{i=0}^{d} \binom{i}{t} = \binom{d+1}{t+1} \leq (d+1)^{t+1}.$$

Consequently, there exists a positive constant c_1 such that the condition

$$d = [c_1 n^2 \operatorname{Log} n]$$

implies $2^{d+1} > (d+1)^{n(n+1)/2}$, and hence, also $\operatorname{Card}(E) > \operatorname{Card}(F)$. Hence (pigeon hole principle!), there exist two distinct polynomials of E which have the same image under f. Their difference P is the polynomial we were looking for.

We have demonstrated the following result.

PROPOSITION 4.9. — *Let n be any positive integer. Then, there are nonzero polynomials P and Q with integer coefficients, such that* $\deg(Q) = n$, *Q divides P and such that the length of Q satisfies*

$$\mathrm{L}(Q) \geq c_2 2^n (n^2 \operatorname{Log} n)^{-1/2} ||P||,$$

where c_2 $(= 1/c_1)$ is an absolute positive constant.

Remark

According to Theorem 4.5 below, the degree d of any polynomial P with integer coefficients of height equal to 1 that is a multiple of $(X-1)^n$ satisfies $d \geq n^2/(8 \operatorname{Log} n)$.

5. The distribution of the roots of a polynomial

This section deals with the study of the distribution of the arguments of the roots of a polynomial with complex coefficients, in terms of the degree of this polynomial and of its length. The results given here show precisely the fact, which is intuitively obvious, that a polynomial with many roots in some direction must have big coefficients [a typical example is $(X-1)^n$].

The first result only deals with real roots, but, in the second one, we consider all the complex roots.

1. An upper bound for the number of real roots

The following theorem, up to the value of the constant, is due to E. Schmidt *. It improves a result of A. Bloch et G. Pólya †. With the constant equal to 4, which is the best possible, this theorem is due to I. Schur ‡.

THEOREM 4.5. — *Let*

$$P(X) = a_n X^n + a_{n-1} X^{n-1} + \cdots + a_0 ,$$

be a polynomial with complex coefficients, where a_0 and a_n are nonzero. We designate by $\mathrm{L}(P)$ the length of P, then, the number r of real roots of P satisfies the inequality

$$r^2 \leq 4n \operatorname{Log}\left(\frac{\mathrm{L}(P)}{\sqrt{|a_0 a_n|}}\right).$$

Proof

We demonstrate this result only with a nonexplicit constant. First, notice that we have to prove only that the number r' of real roots of the polynomial P, which are of absolute value ≥ 1, satisfies

$$r' \leq c\sqrt{n \operatorname{Log}\{L(P)/|a_n|\}}.$$

Indeed, the estimate we are looking for is an immediate consequence of this result applied to the polynomial P and to its reciprocal polynomial.

For any positive integer k, the number r' is at most equal to k increased by the number of real roots of absolute value ≥ 1 of the polynomial

$$F(X) = n^{-k}\left(X\frac{d}{dX}\right)^k P(X),$$

(argue by induction on k, using the theorem of Rolle), and this last number is also the number of real roots of absolute value ≤ 1 of the reciprocal polynomial

$$F^*(X) = X^n F(X^{-1}).$$

---

* *Preuss. Akad. Wiss. Sitzungsber.*, 1932, p. 321.

† On the roots of certain algebraic equations, *Proc. London Math. Soc.*, 33, 1931, p. 102–114.

‡ *Preuss. Akad. Wiss. Sitzungsber.*, 1933, p. 403–428.

We can verify easily that

$$F^*(X) = \sum_{j=0}^{n-k} a_{n-j}\Big(1 - \frac{j}{n}\Big)^k X^j.$$

Let R be a real number, $R > 1$. The inequality of Jensen implies that the total number of (complex) roots of the polynomial F^* in the disc $|z| \leq 1$ is bounded above by the quantity

$$\frac{\max\limits_{|z|=R} \{|F^*(z)|/|F^*(0)|\}}{\operatorname{Log} R}.$$

By choosing $R = e^{k/n}$, we get

$$\max\big\{|F^*(z)|\,;\, |z| = R\big\} \leq \sum_{j=0}^{n} |a_{n-j}|\Big(1 - \frac{j}{n}\Big)^k e^{kj/n} \leq \mathrm{L}(P),$$

the number of zeros of F^* in the disc $|z| \leq 1$ is at most

$$\frac{n}{k} \cdot \operatorname{Log} \frac{\mathrm{L}(P)}{|a_n|}.$$

Hence

$$r' \leq k + \frac{n}{k} \operatorname{Log} \frac{\mathrm{L}(P)}{|a_n|},$$

and the choice

$$k = \left[\left(n \operatorname{Log} \frac{\mathrm{L}(P)}{|a_n|}\right)^{1/2}\right] + 1$$

leads to the result. * □

---

* This demonstration shows that we can take $c = 3$ and thus $C = 6$ is in the statement of the theorem.

2. Distribution of the arguments of the roots of a polynomial

The following result, due to Erdös and Turan †, generalizes the previous theorem.

THEOREM 4.6. — *Let*

$$P(X) = a_n X^n + a_{n-1} X^{n-1} + \cdots + a_0 ,$$

with $a_0 a_n \neq 0$, *be a polynomial with complex coefficients of length* $\mathrm{L}(P)$. *Let* $z_1 = r_1 \exp(i\varphi_1)$, ..., $z_n = r_n \exp(i\varphi_n)$ *be the roots of the polynomial* P, *with* $r_k \in \mathbb{R}$, r_k *positive and* $0 \leq \varphi_k < 2\pi$ *for all indices* $k = 1$, ..., n. *Then the number* $N(\alpha, \beta)$ *of arguments* φ_k *that belong to the real interval* $[\alpha, \beta]$ *satisfies the inequality*

$$\left| N(\alpha, \beta) - \frac{\alpha - \beta}{2\pi} n \right| < 10 \sqrt{n \operatorname{Log} \frac{L(P)}{\sqrt{|a_0 a_n|}}} \, .$$

Remark

Ganelius gave another proof of this result. He obtained the value $c = 2.62$ instead of the constant 10 in the previous statement‡.

6. Separation of the roots of a polynomial

1. Notations

In this section, we consider a polynomial P, of degree d, where d is at least equal to two, with complex coefficients and with complex roots $\alpha_1, \ldots,$ α_d, which is given by

$$P(X) = a_d X^d + \cdots + a_0 = a_d (X - \alpha_1) \cdots (X - \alpha_d), \ a_d a_0 \neq 0 ,$$

and whose roots are numbered so that we have $|\alpha_1| \geq \cdots \geq |\alpha_d|$.

The minimal distance between two distinct roots of P is the quantity

$$\operatorname{sep}(P) := \min \left\{ |\alpha_i - \alpha_j| \, ; \, \alpha_i \neq \alpha_j \right\} ,$$

† On the distribution of the roots of polynomials, *Ann. Math.*, 51, 1950, p. 105–119.

‡ T. Ganelius, Sequences of analytic functions and their zeros, *Arkiv för Math.*, 3, 1958, p. 1–50.

where, by convention, $\text{sep}(P) = \infty$ if P does not have two distinct roots. The discriminant Δ of the polynomial P is given by the formula

$$\Delta = a_d^{2d-2} \prod_{i<j} (\alpha_i - \alpha_j)^2 .$$

If the coefficients a_i are all integers, then the discriminant Δ is also an integer.

2. A lower bound for $\text{sep}(P)$

The useful quantity to consider to get a lower bound for $\text{sep}(P)$ is the discriminant of the polynomial. We know that the discriminant is also given by the expression

$$\Delta = \pm a_d^{2d-2} \Big\{ \det \big(\alpha_k^h\big)_{0<k\leq d,\, 0\leq h<d} \Big\}^2 .$$

With an elementary manipulation of the determinant D, which is in the right-hand side and the inequality of Hadamard, we easily get the inequalities

$$\begin{aligned} |D| &\leq \Big(\sum_{h=1}^{d-1} |\alpha_i^h - \alpha_j^h|^2 \Big)^{1/2} \prod_{k \neq i} \big(1 + |\alpha_k|^2 + \cdots + |\alpha_k|^{2d-2}\big)^{1/2} \\ &< |\alpha_i - \alpha_j| \; d^{(d+2)/2} \max\{1, |\alpha_i|\}^{-1} \Big(\mathrm{M}(P)/|a_d|\Big)^{d-1} , \end{aligned}$$

for any choice of distinct indices i and j. By taking two indices i and j such that $|\alpha_i - \alpha_j| = \text{sep}(P)$, we get the following result.

PROPOSITION 4.10. — *Let P be a polynomial with complex coefficients of discriminant Δ. Then, the minimal distance between two of its distinct roots, denoted* $\text{sep}(P)$, *satisfies the inequality*

$$\text{sep}(P) > d^{-(d+2)/2} \; |\Delta|^{1/2} \; \mathrm{M}(P)^{1-d} \geq d^{-(d+2)/2} \; |\Delta|^{1/2} \; ||P||^{1-d} .$$

The most frequent case for the estimate of $\text{sep}(P)$ is the case of polynomials with integer coefficients. Then Proposition 4.10 can be dramatically simplified.

THEOREM 4.6. — *Let P be a nonconstant polynomial with integer coefficients, then* $\operatorname{sep}(P)$ *satisfies*

$$\operatorname{sep}(P) > d^{-(d+2)/2} \cdot \mathrm{M}(P)^{1-d} \geq d^{-(d+2)/2} \cdot ||P||^{1-d}.$$

Proof

When the polynomial P has no multiple root, we only have to notice that the discriminant of P is a nonzero integer, and hence we have $|\Delta| \geq 1$. However, if P admits multiple roots, there exists a polynomial Q, with integer coefficients, that divides P and for which $\operatorname{sep}(Q) = \operatorname{sep}(P)$. We conclude by applying the estimate of the corollary to the polynomial Q, using the trivial inequalities $\mathrm{M}(Q) \leq \mathrm{M}(P)$ and $\deg(Q) \leq \deg(P)$. □

3. Other lower bounds for the distance between two roots

Let us again study the discriminant Δ.

Let Ω be a set of indices (i,j) such that $\Omega \subset \{(i,j)\,;\, 1 \leq i < j \leq d\}$ and $\Omega' = \{j\,;\,(i,j) \in \Omega\}$. By using the trivial inequality

$$|z - z'| \leq 2 \max\{1, |z|\} \cdot \max\{1, |z'|\},$$

we get the estimate

$$\begin{aligned}
|\Delta| &= |a_d|^{2d-2} \prod_{\{i,j\}\in\Omega} |\alpha_i - \alpha_j|^2 \prod_{\{i,j\}\notin\Omega} |\alpha_i - \alpha_j|^2 \\
&\leq 2^{d(d-1)/2-k} \cdot |a_d|^{2d-2} \prod_{\{i,j\}\in\Omega} |\alpha_i - \alpha_j|^2 \\
&\qquad \times |\alpha_1|^{2(d-1)} \cdot |\alpha_2|^{2(d-2)} \cdots |\alpha_{d-1}|^{-2} \cdot \Pi^{-2},
\end{aligned}$$

where $k = \operatorname{Card}\Omega$ and $\Pi = \prod_{j\in\Omega'} |\alpha_j|$.

Hence we obtain the following result.

PROPOSITION 4.11. — *Let Ω be a set of k couples of indices (i,j) that satisfy the inequalities $1 \leq i < j \leq d$.*

We designate by Ω' the set of indices j such that (i,j) belongs to Ω. We suppose that P is a polynomial with real coefficients. Using the above notations for the polynomial P and its roots, we have

$$\begin{aligned}
\prod_{(i,j)\in\Omega} |\alpha_i - \alpha_j| &\geq |\Delta|^{1/2} \cdot 2^{k-d(d-1)/2} \cdot |a_d|^{1-d} \\
&\qquad \times |\alpha_1|^{1-d} |\alpha_2|^{2-d} \cdots |\alpha_{d-1}|^{-1} \prod_{j\in\Omega'} |\alpha_j|.
\end{aligned}$$

This proposition has the following applications.

COROLLARY. — *The notations are the same as in Proposition 4.11. Let α_i and α_j be two distinct roots of the polynomial P. Then, we have the following estimates :*

(i) If α_i is real and α_j is not, then

$$|\alpha_i - \alpha_j| \geq |\Delta|^{1/6}\ |\alpha_j|\ \left(2^{2-d(d-1)/2}\ |a_d|^{1-d}\ |\alpha_1|^{1-d} \cdots |\alpha_{d-1}|^{-1}\right)^{1/3},$$

which implies

$$|\alpha_i - \alpha_j| \geq \left(|\Delta|^{1/2}\ 2^{2-d(d-1)/2}\ \mathrm{M}(P)^{1-d}\right)^{1/3}.$$

(ii) If α_i and α_j are both imaginary and nonconjugate then

$$|\alpha_i - \alpha_j| \geq |\Delta|^{1/4}\ 2^{1-d(d-1)/4}\ |\alpha_j|\ \left(|a_d|^{1-d}|\alpha_1|^{1-d} \cdots |\alpha_{d-1}|^{-1}\right)^{1/2},$$

which implies

$$|\alpha_i - \alpha_j| \geq |\Delta|^{1/4}\ 2^{1-d(d-1)/4}\ \mathrm{M}(P)^{-(d-1)/2}.$$

Proof

In case (i), take $\Omega = \{(i,j),(i,j'),(j,j')\}$, where j' is the index of the complex conjugate of α_j. Then, use the proposition and the inequality

$$|\alpha_j - \alpha_{j'}| \leq 2|\alpha_i - \alpha_j|.$$

In case (ii), we choose $\Omega = \{(i,j),(i',j')\}$ where i' and j' are the indices of the complex conjugates of α_i and α_j. □

4. Use of Galois properties

Suppose that the polynomial P with integer coefficients is irreducible on the field $\mathbb{Q}$ of rational numbers. Let d be the degree of P and let δ be the degree of the field $\mathbb{Q}(\alpha_i)$ over the field $\mathbb{Q}$. Then consider the number q defined by the formula $q = a_d^{2\delta} \prod_\sigma (\alpha_i - \alpha_j)$, where σ runs over the set E of the embeddings of the field $\mathbb{Q}(\alpha_i, \alpha_j)$ in the field $\mathbb{C}$. This number q is a nonzero rational integer, hence

$$1 \leq |q| \leq |\alpha_i - \alpha_j|\ |a_d|^{2\delta} \prod_{\sigma \neq \mathrm{Id}} |\sigma(\alpha_i) + \sigma(\alpha_j)|.$$

Therefore,

$$1 \leq |\alpha_i - \alpha_j|\ |a_d|^{2\delta} \prod_{\sigma \neq \mathrm{Id}} \left(2 \max\left\{|\sigma(\alpha_i)|,\ |\sigma(\alpha_j)|\right\}\right).$$

Recall that we have assumed $|\alpha_1| \geq |\alpha_2| \geq \cdots \geq |\alpha_n|$. By using the fact that for each α_k the family $\big(\sigma(\alpha_k)\big)_{s\in E}$ contains each root α_h repeated exactly δ times, we get

$$1 \leq |\alpha_i - \alpha_j|\ |a_d|^{2\delta}\ 2^{\delta d-1}\ |\alpha_1^{2\delta}\alpha_2^{2\delta}\cdots\alpha_{\delta'}^{2\delta}\ \alpha_{\delta'+\varepsilon}^{\varepsilon\delta}|,$$

where we have written $\delta' = [\delta/2]$ and $\varepsilon = \delta - 2\delta'$.

Hence, we have the following lower bound for $|\alpha_i - \alpha_j|$:

$$|\alpha_i - \alpha_j| \geq |a_d|^{-2\delta}\ 2^{-\delta d+1}\ |\alpha_1\alpha_2\cdots\alpha_{\delta'}|^{-2\delta}\ |\alpha_{\delta'+\varepsilon}|^{-\varepsilon\delta},$$

which implies the following result, which is not as precise but is simpler,

$$|\alpha_i - \alpha_j| \geq 2^{-\delta d+1}\ \mathrm{M}(P)^{-2\delta}.$$

In general, the degree δ is equal to $d-1$ and the argument given here does not add anything new; nevertheless, if δ is small enough — in fact for $\delta < (d-1)/2$ — the estimate that we have obtained may be better than the previous ones.

5. An example

The aim of this section is to build a monic polynomial with integer coefficients, irreducible over the rationals, and with two very close roots.

Let $d \geq 3$ be a fixed integer and let a be some integer ≥ 10. Then consider the following monic polynomial $P = X^d - 2(aX-1)^2$. The test of irreducibility of Eisenstein (see Exercise 9, Ch. 7) applied for the prime number 2, shows that the polynomial P is irreducible over the ring of rational integers. Now show that P has two roots very closed to the point $1/a$. First, note that we have $P(a^{-1}) = a^{-d} > 0$. But, if $h = a^{-(d+2)/2}$, we have

$$P(a^{-1} \pm h) < 2a^{-d} - 2a^2h^2 = 0.$$

Thus, the polynomial P has two real roots contained in the open real interval $\left]a^{-1} - h, a^{-1} + h\right[$, which proves the inequality

$$\operatorname{sep}(P) < 2h = 2a^{-(d+2)/2}.$$

This construction† implies the following result.

PROPOSITION 4.12. — *For any integer $d \geq 3$, there exist an infinity of monic polynomials P, with integer coefficients, irreducible on the rational numbers and satisfying*

$$\operatorname{sep}(P) < 2^d \, \mathrm{H}(P)^{-(d+2)/4}.$$

† This construction has been published in the author's article "Some Inequalities about univariate Polynomials", *Proceed.* 1981 A.C.M. *Symp. on Symbolic and Algebraic Computation*, Snowbird, Utah.

Exercises

1. The aim of this exercise is to demonstrate the theorem of Gauss-Lucas : When f is a polynomial with complex coefficients, the set of the complex zeros of its derivative f' is contained in the convex hull of the set of the complex zeros of f.

1°) Let P be a polynomial with complex coefficients whose roots are all in the half plane $\mathrm{Im}(z) \geq 0$. Demonstrate that the roots of the derivative of this polynomial belong to this same half plane.

[Hint : Let $z_1, \ldots, z_d$ be these roots, with $\mathrm{Im}(z_j) \geq 0$ for $1 \leq j \leq d$ then

$$\frac{P'(z)}{P(z)} = \sum_{j=1}^{d} \frac{1}{z - z_j} = \sum_{j=1}^{d} \frac{\bar{z} - \bar{z}_j}{|z - z_j|^2}$$

and

$$\mathrm{Im}(z) < 0 \Longrightarrow \mathrm{Im}\left(\frac{1}{z - z_j}\right) > 0,$$

hence $P'(z) \neq 0$ if z satisfies $\mathrm{Im}(z) < 0$.]

2°) Deduce the theorem of Gauss-Lucas.

2. Let $P(X) = X^n + a_m X^m + \cdots + a_n$, where $m < n$, be a polynomial with complex coefficients and let A be the maximum of the modulus of the a_j, for $m \leq j \leq n$. Demonstrate that any complex root z of this polynomial satisfies the inequality $|z| < 1 + A^{1/m}$.

3. Let P and Q be two polynomials with complex coefficients, of degrees p and q, respectively. Demonstrate that their resultant satisfies the inequalities

(i) $$|\mathrm{Res}(P, Q)| \leq ||P||^q \; ||Q||^p$$

and

(ii) $$|\mathrm{Res}(P, Q)| \leq \mathrm{M}(P)^q \, \mathrm{L}(Q)^p .$$

[For (i), use the determinant of Sylvester and the inequality of Hadamard. For (ii), if $x_1, \ldots, x_p$ are the roots of P and c is its leading coefficient, use the relation

$$\operatorname{Res}(P, Q) = \pm c^q Q(x_1) \cdots Q(x_p).]$$

4. Let f be a monic polynomial with complex coefficients, of measure M, whose roots are $z_1, \ldots, z_n$. Demonstrate the inequalities

$$\mathrm{L}(f) \leq \prod_{k=1}^{n} (1 + |z_k|) \leq 2^n M.$$

5. Let α and β be two non-zero algebraic numbers of respective degrees equal to m and n. Demonstrate that the numbers α/β and $\alpha + \beta$ are algebraic numbers of degree at most mn [use Exercise 25, Ch. 3] and that their measures satisfy the respective inequalities

$$\mathrm{M}(\alpha/\beta) \leq \mathrm{M}(\alpha)^n \cdot \mathrm{M}(\beta)^m$$

and

$$\mathrm{M}(\alpha + \beta) \leq 2^{mn}\, \mathrm{M}(\alpha)^n\, \mathrm{M}(\beta)^m.$$

[To demonstrate the second inequality, note that

$$\max\{1, x + y\} \leq 2 \max\{1, x\} \cdot \max\{1, y\},$$

for positive reals numbers x and y.]

6. Let P be a squarefree polynomial with integer coefficients of degree $n \geq 2$. Let α be any complex root of the derivative of P. Demonstrate the inequality

$$|P'(\alpha)| \geq n^{-n}\, \mathrm{L}(P)^{-2n+1}.$$

[Consider the discriminant Δ of P. Let $\alpha_1 = \alpha, \ldots, \alpha_{n-1}$ be the roots of P' and let a be the leading coefficient of P. Show that $|\Delta| \geq 1$ and that

$$\prod_{i=2}^{n} |P(\alpha_i)| \leq \mathrm{L}(P)^{n-1}\, \mathrm{M}(P')^n \leq n^n\, \mathrm{L}(P)^{2n-1}.]$$

7. Let P be a polynomial of degree $n \geq 2$ and let a be a complex number. We write $T(x) = P(a) + P'(a)(x-a)$. Demonstrate that, for $|x-a| \leq 1/2$, and x complex, we have

$$|P(x) - T(x)| \leq \sum_{i=2}^{n} \binom{n}{i} \mathrm{L}(P) \max\{1, |a|\}^{n-2} |x-a|^2 \, 2^{-i+2}.$$

[Use Taylor formula.]

Deduce the inequality

$$|P(x) - T(x)| \leq 4\,(3/2)^n \, \mathrm{L}(P) \max\{1, |a|\}^{n-2} |x-a|^2.$$

8. Let P be a polynomial with complex coefficients and let α be one of its roots. We write $Q(X) = P(X)/(X-\alpha)$. Demonstrate that the height of the polynomial Q satisfies $\mathrm{H}(Q) \leq \deg(P) \cdot \mathrm{H}(P)$.

[Use the formulas of the euclidian division.]

9. Let A be the set of complex numbers that are algebraic.

1°) Show that A is countable.

2°) Demonstrate that A is a subfield of $\mathbb{C}$ and that it is an algebraically closed field.

10. Let α be an algebraic number of degree n and let Q be a polynomial with integer coefficients which is not zero at the point α.

1°) Demonstrate the inequality

$$|Q(\alpha)| \geq \mathrm{L}(Q)^{-n+1} \, \mathrm{H}(\alpha)^{-\deg(Q)}.$$

2°) Deduce the following result : if P is a squarefree polynomial with integer coefficients of degree n and if α is a root of P, then we have

$$|P'(\alpha)| \geq \big(n\,\mathrm{L}(P)\ \mathrm{M}(P)\big)^{-n+1}.$$

3°) We come back to the first question, in the case where the polynomial Q is of degree one, let us say $Q(X) = qX - p$. Demonstrate that there exists a positive constant $C = C(\alpha)$ such that

$$\left|\alpha - \frac{p}{q}\right| \geq C q^{-n} \quad \text{if} \quad p/q \neq \alpha.$$

This inequality is called *Liouville inequality*.

An irrational number α such that

$$\liminf_{p,q\in\mathbb{Z}} \frac{\operatorname{Log}\left|\alpha - \frac{p}{q}\right|}{\operatorname{Log} q} = -\infty$$

is called a *Liouville number*; according to the previous study, such a number is *transcendental* (that is, nonalgebraic). Demonstrate that the number

$$\sum_{n=0}^{\infty} 2^{-n!}$$

is a Liouville number.

More generally, let (ε_n) be a sequence of integers equal to 0 or 1 containing an infinity of 1. Demonstrate that the number

$$\sum_{n=0}^{\infty} \varepsilon_n \, 2^{-n!}$$

is also a Liouville number. Deduce that the set of Liouville numbers has the power of the continuum (compare with the previous exercise).

11. Let P be a polynomial with complex coefficients and with complex roots $\alpha_1, \ldots, \alpha_n$. Let ξ be a complex number and let α be a root of P at a minimal distance of the point ξ. Demonstrate the inequality

$$|\xi - \alpha| \leq \frac{2^{n-1}\,|P(\xi)|}{|P'(\alpha)|}.$$

More generally, demonstrate that we have

$$|\xi - \alpha|^r \leq |\alpha - \alpha_1| \cdots |\alpha - \alpha_{r-1}| \, \frac{2^{n-r}\,|P(\xi)|}{|P'(\alpha)|},$$

for r integer, $1 \leq r \leq n$, and $\alpha \notin \{\alpha_1, \ldots, \alpha_{r-1}\}$.

12. Let P be a squarefree polynomial with integer coefficients, with roots $\alpha_1, \ldots, \alpha_n$, and of leading coefficient equal to a. We call ρ the maximal value of $|\alpha_i|$, for the indices $i = 1, \ldots, n$, and we write

$$s = \operatorname{sep}(P) = \min\left\{|\alpha_i - \alpha_j|\,;\, \alpha_i \neq \alpha_j\right\}.$$

If α is any root of P, demonstrate the inequalities

$$\left\{\mathrm{L}(P')\,\mathrm{M}(\alpha)\right\}^{-n+1} \leq |P'(\alpha)| \leq a\,s\,(2\rho)^{n-2}.$$

13. Let P and Q be two linear polynomials (linear means of degree one) with complex coefficients. Demonstrate the inequality

$$\frac{\sqrt{5}-1}{2}\,\mathrm{H}(P)\cdot\mathrm{H}(Q)\leq\mathrm{H}(P\cdot Q),$$

and determine the cases where equality occurs.

14. With the results of this chapter, demonstrate that there exists an algorithm that solves the following problem :

> Given a polynomial P with integer coefficients and with distinct complex roots $\alpha_1, \ldots, \alpha_m$ and ε a fixed positive number; determine m complex numbers $z_1, \ldots, z_m$ which belong to $\mathbb{Q}(i)$ such that $|\alpha_j - z_j| < \varepsilon$ for all the indices $j = 1, 2, \ldots, m$.

[It is mainly sufficient to use the bounds on the roots and a lower bound for the distance $\mathrm{sep}(P)$.]

Moreover, demonstrate that there exists an algorithm that determines the number of real roots of P and a rational approximation of each of them up to ε.

[Give a lower bound for $|\mathrm{Im}(\alpha_j)|$ when α_j is imaginary, using the obvious formula $\mathrm{Im}(\alpha_j) = (\alpha_j - \bar{\alpha}_j)/2$.]

Generalize both previous results to the case of the polynomials of $\mathbb{C}[X]$ with algebraic coefficients.

15. Demonstrate that there exists an algorithm to solve the following problem :

> Given a nonconstant polynomial P with integer coefficients, determine polynomials $P_1, \ldots, P_m$ with integer coefficients, irreducible over $\mathbb{Z}$ and such that $P = P_1 \cdots P_m$.

[Just use bounds for the roots. See Chapter 7.]

Generalize this result, in a suitable way, to the case of polynomials with algebraic coefficients.

16. Demonstrate that there exist at most $n(2H+1)^n$ algebraic numbers of degree at most n and of height at most H (the height of an algebraic number is, by definition, the height of its minimal polynomial over $\mathbb{Z}$). Deduce the following theorem of Kronecker : a complex number α, which is an algebraic integer over $\mathbb{Z}$ and whose conjugates (including itself) all are of modulus at most 1, is necessarily a root of unity.

[Demonstrate that the set of powers of α is finite.]

Show that the theorem of Kronecker implies the following assertion : for any integer $d > 1$, there exists a constant $C = C(d) > 1$ such that any algebraic integer whose conjugates (including itself) all are of modulus at most C is necessarily a root of unity.

17. Some inequalities about multivariate polynomials

1°) Let P be a polynomial with complex coefficients and with complex roots $\alpha_1, \ldots, \alpha_n$ all different from zero. We write $\alpha_j = |\alpha_j|\zeta_j$ for all $j = 1, \ldots, n$. Demonstrate the following implication :

$$x \in \mathbb{C} \text{ and } |x| = 1 \Longrightarrow \max\{|P(z)|\,;\,|z| = 1\} \cdot \prod_{j=1}^{n} |x - \zeta_j| \leq 2^n\,|P(x)|.$$

[Demonstrate that $(1 + |\alpha_j|)(x - |\zeta_j|) \leq 2|x - \alpha_j|$ if $|x| = 1$.]

2°) Let $P \in \mathbb{C}[X_1, \ldots, X_n]$ be a multivariate polynomial such that $P = P_1 \cdots P_m$. Using the previous question, demonstrate the inequality

$$\prod_{j=1}^{m} ||P_j||_2 \leq 2^{D-s/2}\,||P||_2.$$

[Use the formula of Parseval several times, by integrating with respect to each of the variables. The proof, due to Gel'fond, can be found in the book of Stolarsky*.]

3°) In the same way, for any positive integer k, demonstrate the following inequality :

$$||P||_2{}^k \leq ||P^k||_2 \prod_{h=1}^{n} (1 + 2kd_h),$$

* K. B. Stolarsky, *Algebraic Numbers and Diophantine Approximation*, Marcel Dekker, New York, 1974, Ch. 4, Sec. 2.

where, as usual, d_h is the partial degree of the polynomial P relative to the variable X_h.

[See again Stolarsky, Chapter 4, Section 2.]

18. Let m be an integer, $m \geq 2$, and put $n = m^3$. Then, consider the polynomial $P(X) = X^n - 1$, and — for $1 \leq k \leq m$ — write

$$P_k = \prod_{j=1}^{h-1} \left(X - \exp\left\{\frac{2i\pi(k-1)h + j}{n}\right\}\right), \quad \text{where} \quad h = m^2.$$

Demonstrate that the relation $P = P_1 \cdots P_m$ holds and that there exist strictly positive constants c and c' such that we have

$$\prod_{j=1}^{m} ||P_j|| \geq c\, 2^{n - c' n^{1/3} \operatorname{Log} n}.$$

[This result is also due to Gel'fond, (see Stolarsky, p. 55); it can be compared to the one given here in Section 4.5.]

19. Let n be a fixed positive integer. Demonstrate that there exist strictly positive constants $c = c(n)$ and $C = C(n)$, such that for any polynomial P of degree n and of height H we have the inequalities

$$c\,H \leq \max\{|P(k)|\,;\, k = 0, 1, \ldots, n\} \leq C\,H.$$

[The inequality on the right is trivial. Whereas the one on the left can be demonstrated by using the formulas giving the polynomial of Lagrange, see Exercise 5 of Chapter 3.]

Demonstrate a similar result for the maximum of the values $|P(\zeta)|$, when the number ζ runs over the set E of the nth roots of unity.

[To demonstrate the left inequality, we can, this time, use the formula

$$P(0) = \sum_{\zeta \in E} P(\zeta),$$

and its similars which give the various coefficients of the polynomial P.]

20. Some estimates for symmetric functions of the roots of a polynomial

1°) Let $|z_1|, \ldots, |z_n|$ be complex numbers of modulus 1. Demonstrate the inequality

$$\prod_{1\leq i<j\leq n} |z_i - z_j| \leq n^{n/2}.$$

[Use the formula of Vandermonde as well as the inequality of Hadamard.]

2°) Deduce that if $|z_1|, \ldots, |z_n|$ are n complex numbers of modulus $\rho \geq 1$ then, we have the inequality

$$\prod_{1\leq i<j\leq n} |z_i z_j - 1| \leq \rho^{n(n-1)}\, n^{n/2}.$$

[Consider the product

$$\prod_{1\leq i<j\leq n} |z_i z_j - \omega_i\omega_j| \quad \text{for} \quad |z_1|, \ldots, |z_n|, |\omega_1|, \ldots, |\omega_n| \leq \rho.$$

Using the principle of the maximum (see the following exercise), demonstrate that the maximum of this product is reached for values of the numbers $|z_1|, \ldots, |z_n|, |\omega_1|, \ldots, |\omega_n|$, which all are of a modulus equal to ρ. Deduce that this product is bounded above by the expression

$$\prod_{1\leq i<j\leq n} |\omega_i\omega_j| \cdot \max\Big\{ \prod_{1\leq i<j\leq n} |u_i\bar{u}_j - 1|\,;\, |u_1| = \ldots = |u_n| = 1\Big\}.$$

Then, conclude by using the previous question.]

21. Principle of the maximum (for polynomials)

Let P be a nonconstant polynomial with complex coefficients and let ρ be a positive real number. We call M the upper bound of $|P(z)|$ when z runs over $D = \{z\,;\, |z| \leq \rho\}$.

1°) Demonstrate that there exists at least one point x of this disc for which we have $|P(x)| = M$ and that such a point x is necessarily on the circle $|z| = \rho$.

[For the second assertion, suppose the contrary : if x satisfies $|x| < \rho$, considering the value $P(x+y)$, for y small, demonstrate that there exists one point $x+y$ of D such that we have $|P(x+y)| > |P(x)|$.]

2°) Suppose, however, that P is never zero on the disc D. Call m the upper bound of $|P(z)|$ when z runs over this disc. Demonstrate that there exists at least one point x of the disc D for which we have $|P(x)| = m$ and that moreover, such a point x is necessarily on the circle $|z| = \rho$.

[Use the same technique as in the first question.]

Deduce another demonstration of the theorem of d'Alembert.

22. Consider again Exercise 34 of Chapter 3, in which we have shown that, if P and Q are two polynomials with coefficients in a ring A, then there exist two polynomials U and V of $A[X]$ such that the resultant R of the polynomials P and Q is given by

$$(*) \qquad R = U(X)P(X) + V(X)Q(X),$$

with $\deg(U) < \deg(Q)$ and $\deg(V) < \deg(P)$.

1°) Suppose that P and Q have coefficients in a subring of $\mathbb{C}$, demonstrate that we can find polynomials U and V as above and that satisfy, moreover

$$\mathrm{H}(U) \leq ||P||^{q-1}\, ||Q||^{p} \quad \text{and} \quad \mathrm{H}(V) \leq ||P||^{q}\, ||Q||^{p-1},$$

where $p = \deg(P)$ and $q = \deg(Q)$.

2°) Deduce that if the polynomials P and Q have integer coefficients and have no common factor, for any complex number z, we have

$$|P(z)| + |Q(z)| \geq \left(\max\{p,q\}\right)^{-1} ||P||^{-q}\, ||Q||^{-p}.$$

[Use relation $(*)$ for $X = z$ and the fact that we have $|R| \geq 1$.]

23. Let $P \in \mathbb{Z}[X]$ be a nonconstant polynomial of degree n, of measure M, and let z be a complex number for which we have $|P(z)| \leq 1$. We want to demonstrate that there exists a polynomial Q with integer coefficients, *primary factor* (in a factorial ring, an element is said to be *primary* if it is a power of an irreducible element) of P such that

$$|Q(z)| \leq \left(2(n-1)\right)^{-2} \left\{ \binom{n}{n'} M \right\}^{-2(n-1)},$$

where $n' = [n/2]$.

This is the procedure. Let $P_1, \ldots, P_k$ be the primary factors of the polynomial P, and let us write $y_j = |P_j(z)|$ for $j = 1, 2, \ldots, k$. Notice that there exists an index h, $1 \le h \le k$, such that

$$y_1 \cdots y_{h-1} \le y_h \cdots y_k \quad \text{and} \quad y_1 \cdots y_h \le y_{h+1} \cdots y_k .$$

Using the exercise above, demonstrate the inequality

$$\min\{y_1 \cdots y_{h-1},\, y_{h+1} \cdots y_k\} \ge \frac{1}{2(n-1)} \binom{n}{n'}^{-n+1} M^{-n+1}.$$

24. Let $\alpha_1, \ldots, \alpha_n$ be algebraic numbers of respective measures equal to $M_1, \ldots, M_n$ and of respective degrees $d_1, \ldots, d_n$. Let $b_1, \ldots, b_n$ be rational integers such that the number

$$\Lambda = b_1 \log \alpha_1 + \cdots + b_n \log \alpha_n$$

(where $\log \alpha_i$ is any determination of the logarithm of α_i, for $1 \le i \le n$) is nonzero, then demonstrate that this number satisfies the inequality

$$|\Lambda| > 2^{-D} \exp\Big\{-D\big(|b_1|h_1 + \cdots + |b_n|h_n\big)\Big\},$$

where D is the degree of the field $\mathbb{Q}(\alpha_1, \ldots, \alpha_n)$ over the field $\mathbb{Q}$ of rational numbers and where we have written $h_i = (\operatorname{Log} M_i)/d_i$ for $i = 1, 2, \ldots, n$. [Suppose $|\Lambda| \le 1/2$. Then $\Lambda \in 2\pi\mathbb{Z}$ and the number $\zeta = \alpha_1^{b_1} \cdots \alpha_n^{b_n} - 1$ is nonzero.

Moreover, without loss of generality, we can also suppose $b_j \ge 0$ for the indices $j = 1, 2, \ldots, r$ and $b_j \le 0$ for $j = r, r+1, \ldots, n$. Let $a_1, a_2, \ldots, a_n$ be the respective leading coefficients of the minimal polynomials of $\alpha_1, \alpha_2, \ldots, \alpha_n$. Then, show that we have

$$a_1^{|b_1|D/d_1} \cdots a_n^{|b_n|D/d_n} \prod_{\sigma} \Big|\sigma\Big(\zeta\, \alpha_{r+1}^{-b_{r+1}} \cdots \alpha_n^{-b_n}\Big)\Big| \ge 1,$$

where σ runs over the set of the embeddings of the field $\mathbb{Q}(\alpha_1, \ldots, \alpha_n)$ in the field $\mathbb{C}$; which gives

$$|\zeta| \ge 2^{-D+1} \exp\Big\{-D\big(h_1|b_1| + \cdots + h_n|b_n|\big)\Big\}$$

The result follows, using the inequality $|e^z - 1| < 2|z|$, which is true for $0 < |z| \le 1/2$.]

When at least one of the numbers α_j is not real, demonstrate that Λ satisfies the inequality

$$|\Lambda| > 2^{-D'} \exp\Big\{-D'\big(|b_1|h_1 + \cdots + |b_n|h_n\big)\Big\},$$

where $D' = D/2$.

[Consider the product $\zeta\bar{\zeta}$, where ζ is as above and where $\bar{\zeta}$ is the complex conjugate of ζ.]

25. Let K be an algebraically closed field and let F be any nonconstant polynomial in $K[X,Y]$. Prove that the set of the zeros of F in K^2 is infinite. Is this result true if K is the field of real numbers? [Consider the polynomial $X^2 + Y^2 + 1$.]

26. Let P be a monic polynomial of degree d, with complex coefficients given by

$$P(X) = X^d + a_{d-1}X^{d-1} + \cdots + a_0 = (X - \alpha_1)\cdots(X - \alpha_d)$$

with $\rho := |\alpha_1| \ge |\alpha_2| \ge \cdots \ge |\alpha_d|$. Prove that if $1 \le n \le d$ and the polynomial Q is defined by the relation

$$P(X) = (X - \alpha_1)\cdots(X - \alpha_n)\,Q(X),$$

then we have the inequality

$$\big||\alpha_1| - 1\big|\cdots\big||\alpha_n| - 1\big| \cdot \mathrm{H}(Q) < \mathrm{H}(P).$$

[It is sufficient to prove this inequality for $n = 1$. Put

$$Q(X) = X^{d-1} + \cdots + b_0 .$$

Suppose that $|\alpha_1| \ge 1$ and let i be minimal such that $\mathrm{H}(Q) = |b_i|$. The relation $a_i = b_{i-1} - \alpha_1 b_i$ (where $b_{-1} = b_d = 0$) implies

$$\mathrm{H}(P) \ge |a_i| \ge |\alpha_1|\,\mathrm{H}(Q) - |b_{i-1}| > (|\alpha_1| - 1)\,\mathrm{H}(Q).$$

The proof is similar when $|\alpha_1| < 1$.]

Deduce the following :

1°) Suppose that $|\alpha_1| = \cdots = |\alpha_k| = \rho$, then $\rho < H^{1/k} + 1$.

2°) (Cauchy)† We have $\rho < \mathrm{H}(P) + 1$.

3°) Let P be a monic polynomial with real coefficients, and let α be a nonreal root of P, then we have‡ $|\alpha| < \mathrm{H}(P)^{1/2} + 1$.

Remark

One can give many variants of the previous result. For example, if P is equal to the product of R and Q, demonstrate that

$$\mathrm{H}(P) \geq \big(2\,\mathrm{H}(R) - \mathrm{L}(R)\big) \cdot \mathrm{H}(Q).$$

In particular, if α is a nonreal root of a monic polynomial with real coefficients P, $\alpha = |\alpha|\, e^{i\theta}$, prove that $|\alpha| \leq |\cos\theta| + \big\{\mathrm{H}(P) + 1 + \cos^2\theta\big\}^{1/2}$, which is better than Corollary 4.3 if $\cos\theta$ is small enough.

27. Consider a polynomial with complex coefficients

$$f(X) = X^k + a_{k-1}\, X^{k-1} + \cdots + a_0\,.$$

The question we wish to solve is : find a positive real number R such that all roots of f have a modulus less or equal to R. In other words, if ρ is the largest modulus of the roots of the polynomial f, we are looking for an upper bound of ρ.

1°) Let $C(f)$ be the unique positive real root of the polynomial

$$f^*(X) = X^k - |a_{k-1}|\, X^{k-1} - \cdots - |a_0|\,.$$

Verify that $f^*(x) \geq 2x^k - (x + \rho)^k$, for any positive real x. Then, prove that $C(f)$ satisfies

$$(*) \qquad \rho \leq C(f) \leq \rho\,(2^{1/k} - 1)^{-1},$$

Show that both inequalities are sharp.

---

† A.L. Cauchy, *Exercices de Mathématiques*, Quatrième Année, De Bure Frères, Paris, 1829. *Oeuvres*, Ser. II, Vol. IX, Gauthier-Villars, Paris, 1891.
‡ M. Mignotte, An inequality on the greatest roots of a polynomial, *Elemente der Mathematik*, 1991.

[The left inequality is an equality when $f = f^*$, whereas the right inequality is an equality when $f(x) = (x+\rho)^k$.]

2°) Demonstrate that

$$(**) \quad \rho \leq K(f) := 2 \max\{|a_{k-1}|, |a_{k-2}|^{1/2}, |a_{k-3}|^{1/3}, \cdots, |a_0|^{1/k}\},$$

Prove that $K(f) \leq 2k\rho$. [Use the inequalities $\binom{k}{i}^{1/i} \leq k$, for $1 \leq i \leq k$.]

3°) Recall that, if we apply Graeffe's method n times to f, we obtain a polynomial f_n, whose roots are the 2^nth powers of the roots of the polynomial f. Prove that

$$\rho \leq C(f_n)^{2^{-n}} \leq (k/\operatorname{Log} 2)^{2^{-n}} \rho.$$

[Apply $(*)$ to the polynomial f_n.]

Prove that last term tends rapidly to ρ as n increases. In a similar way, show that

$$\rho \leq K(f_n)^{2^{-n}} \leq (2k)^{2^{-n}} \rho.$$

and that last term tends rapidly to ρ as n increases.

[Use inequality $(**)$ instead of $(*)$.]

4°) Consider the polynomial

$$P(X) = X^6 + X^5 + 6X^4 - 5X^3 + 3X^2 + 2.$$

Verify that

$$P_3(X) = X^6 + 4591X^5 + 6056224X^4 - 3116689X^3 + 680865X^2 + 22240X + 256,$$

and prove that $\rho < 2.9$. ¶

¶ Reference for this exercise : J.H. Davenport and M. Mignotte, On finding the largest root of a polynomial, *MAN*, 24, 6, 1990, p. 693–696.

28. Let $P(X,Y)$ be a nonconstant irreducible polynomial over $\mathbb{C}$ (or over any algebraically closed field of characteristic zero). Suppose that there exist two infinite subsets of $\mathbb{C}$, I and J, and two positive integers p and q such that, for x in I, the polynomial $P(x,Y)$ has exactly q roots, and $P(X,y)$ has exactly p roots for y in J. Demonstrate that $\deg_X P = p$ and $\deg_Y P = q$.

[Consider the polynomial $\Delta(Y) = \operatorname{Res}_X(P, P'_X)$. Prove that this polynomial is nonzero. Show that there exists $y \in J$ such that $\Delta(y) \neq 0$. Conclude that $\deg_X P = p$. Same proof for $\deg_Y P$.]

29. Let k and d be two positive integers, $2 \leq k < d$. Prove that the polynomial $X^d - 2\,(aX-1)^k$ has k roots very close to the point $1/a$. [Use the theorem of Rouché.]

Deduce that proposition 4.11 cannot be much improved. [Notice that this result generalizes Proposition 4.12.]

Chapter 5

POLYNOMIALS WITH REAL COEFFICIENTS

In this chapter, we call $\mathbb{R}$ an ordered field such that the field $\mathbb{C} = \mathbb{R}[\sqrt{-1}]$ is algebraically closed. Usually, we consider that $\mathbb{R}$ is the field of real numbers and that $\mathbb{C}$ is the field of complex numbers. We study algorithms to separate the real roots of polynomials.

1. Polynomials irreducible over $\mathbb{R}$

Here, we demonstrate Proposition 4.1.

PROPOSITION 5.1. — *The polynomials with coefficients in $\mathbb{R}$ and irreducible over $\mathbb{R}$ are first-degree polynomials or second-degree polynomials with a strictly negative discriminant.*

Proof

Let P be a monic polynomial, with real coefficients, that is irreducible over the field $\mathbb{R}$ (thus, by definition, P is non constant). If P admits a real root α, then $X - \alpha$ is a polynomial with real coefficients that divides P, hence, since P is irreducible, $P = X - \alpha$. If P has no real roots then it admits at least two imaginary conjugate roots $a + ib$ and $a - ib$, where a and b are real and $b \neq 0$. Then, the polynomial

$$(X - a - ib)\,(X - a + ib) = X^2 - 2aX + a^2 + b^2$$

has real coefficients and divides P, hence it is equal to P. And the discriminant of P is $\operatorname{Discr}(P) = -b^2 < 0$.

Inversely, it is clear that any polynomial with real coefficients of degree one or of degree two and of negative discriminant is irreducible over $\mathbb{R}$. (Indeed, since $\mathbb{R}$ is an ordered field, its negative elements are not squares. Prove this as an exercise.) $\square$

Complement : If P is a quadratic polynomial with real coefficients, irreducible over the field $\mathbb{R}$, then $P(x)$ has constant sign when x runs over the set of real numbers.

2. The theorem of Rolle

1. The theorem of intermediate values

This theorem is a consequence of the following result.

THEOREM 5.1. — *Let f be a polynomial with real coefficients, and let a and b be two real numbers, where $a < b$, such that the values $f(a)$ and $f(b)$ are nonzero. Then the numbers of roots of f in the open interval $]a, b[$, counted with their multiciplities, is even or odd whether the product $f(a)f(b)$ is positive or negative.*

Proof

Let us write

$$f(X) = c \prod_{i=1}^{m}(X - \alpha_i) \prod_{j=1}^{n}(X^2 + \beta_j X + \gamma_j),$$

where each of the polynomials on the right-hand side is irreducible over $\mathbb{R}$. As we have just seen it, each of the quadratic polynomials in this decomposition has positive values at any real point. Suppose that $\alpha_1, \ldots, \alpha_k$ are exactly the roots of polynomial f lying between the points a and b. Then, we have

$$\frac{f(a)}{f(b)} = \frac{a - \alpha_1}{b - \alpha_1} \cdots \frac{a - \alpha_k}{b - \alpha_k} \prod_{i=k+1}^{m} \frac{a - \alpha_i}{b - \alpha_i} \prod_{j=1}^{n} \frac{a^2 + \beta_j a + \gamma_j}{b^2 + \beta_j b + \gamma_j},$$

which immediately leads to the relation

$$\operatorname{sign}\{f(a)f(b)\} = (-1)^k.$$

Hence we have the result. □

The theorem of intermediate values is the following corollary.

COROLLARY. — *Let f be a polynomial with real coefficients, and let a and b be two real numbers, where $a < b$, such that we have $f(a)\, f(b) < 0$. Then the polynomial f has at least one root in the open interval $]a, b[$.*

2. *The theorem of Rolle*

The theorem of Rolle is shown in the following theorem.

THEOREM 5.2 (Rolle). — *Let f be a polynomial with real coefficients and let a and b be two real successive roots of f, with $a < b$. Then the derivative of f has an odd number of roots in the real interval $]a, b[$.*

Proof

Let m be the order of a, then the polynomial f is written

$$f(X) = (X - a)^m g(X), \quad \text{with} \quad g(a) \neq 0,$$

hence

$$f'(X) = m(X - a)^{m-1} g(X) + (X - a)^m g'(X).$$

For a real number h, which is non zero and small enough, we have the estimate

$$f(a + h)/f'(a + h) \approx h/m.$$

This implies the relation

$$\text{sign}\{f(x)f'(x)\} > 0 \quad \text{for} \quad x - a > 0, \text{ small enough.}$$

In the same way

$$\text{sign}\{f(x)f'(x)\} < 0 \quad \text{for} \quad x - b < 0, \text{ small enough.}$$

Because f never takes the value zero on the open interval $]a, b[$ (since a and b are two successive roots of f), by the corollary of Theorem 5.1, the polynomial f does not change of sign on the interval $]a, b[$. Hence, for any real $h > 0$ small enough,

$$f'(a + h)f'(b - h) < 0.$$

We reach the conclusion via theorem 5.1. □

COROLLARY 1. — *Let f be a polynomial with real coefficients that has m roots (counted with their orders of multiplicity) in a closed interval of $\mathbb{R}$. Then its derivative has at least $m-1$ roots in this interval.*

Proof

Prove the corollary as an exercise. □

COROLLARY 5.2. — *Let f be a polynomial with real coefficients, and let a and b be two real numbers, with $a < b$. Then there exists at least one real number c, with $a < c < b$, that satisfies the relation*

$$f(b) - f(a) = (b-a)f'(c).$$

Proof

Apply the theorem to the polynomial

$$u(X) = (b-a)\{f(X) - f(a)\} - (X-a)\{f(b) - f(a)\}.$$

The result is immediate . □

COROLLARY 3. — *Let f be a polynomial with real coefficients, and let a and b be two real numbers, with $a < b$. Suppose that the derivative of f takes a positive value (and respectively a negative value) at each point x of the interval $]a,b[$. Then the function $f(x)$ is strictly increasing (or respectively strictly decreasing) on the interval $]a,b[$.*

Proof

The result is an immediate consequence of Corollary 2. □

3. Estimates of real roots

Most of the results of this section are variants of the estimates of complex roots that were proved in Section 4.2; when it is the case, the new demonstrations will be quite short.

1. The rule of Newton

PROPOSITION 5.2. — *Let f be a polynomial of degree n with real coefficients. Let L be a real number such that we have $f^{(i)}(L) \geq 0$ for $i = 0, 1, \ldots, n$. Then any real root x of the polynomial f satisfies $x \leq L$.*

Proof

Let $x = L+y$ be a real number, where y is positive. By Taylor's formula, we have

$$f(x) = \sum_{i=0}^{n} \frac{y^i}{i!} f^{(i)}(L),$$

hence $f(x)$ is strictly positive for $x > L$. ∎

2. The rule of Lagrange and MacLaurin

PROPOSITION 5.3. — *Let f be a polynomial with real coefficients,*

$$f(X) = X^n + a_1 X^{n-1} + \cdots + a_n,$$

with $a_i \geq 0$ for $i = 1, \ldots, m-1$, and let $A = \max\{-a_m, \ldots, -a_n, 0\}$. Then any real root x of the polynomial f satisfies

$$x < 1 + A^{1/m}.$$

Proof

For $x \geq 1 + A^{1/m}$, we have the relation

$$f(x) \geq x^n - A(1 + x + \cdots + x^{n-m}) > 0.$$

Hence, we have the result. ∎

3. A special case of the rule of Descartes

PROPOSITION 5.4. — *Let f be a polynomial with real coefficients like*

$$f(X) = X^n + a_1 X^{n-1} + \cdots + a_m X^{n-m} - a_{m+1} X^{n-m-1} - \cdots - a_n,$$

with $a_i \geq 0$ for $i = 1, \ldots, n$. If c is a nonnegative real number for which $f(c)$ is also nonnegative, then for any real number $x > c$ we have the inequality $f(x) > 0$. Thus, in such a case, any real root x of the polynomial f satisfies $x \leq c$.

Proof

This is another formulation of Lemma 4.1. □

4. The rule of Cauchy

PROPOSITION 5.5. — *Let a_m, $a_{m'}$, ... with $m > m' > \cdots$ be the strictly negative coefficients of a polynomial f with real coefficients,*

$$f(X) = X^n + a_1 X^{n-1} + \cdots + a_n ,$$

and let k be number of these negative coefficients. Then any real root x of the polynomial f satisfies

$$x \leq \max \left\{ \left(k\,|a_m|\right)^{1/m},\ \left(k\,|a_{m'}|\right)^{1/m'}, \ldots \right\}.$$

Proof

Let M be the value of the maximum considered in the statement of the proposition. For $x > M$, we have

$$f(x) \geq x^n - (a_m\, x^{n-m} + a_{m'}\, x^{n-m'} + \cdots) > 0 .$$

[See again the demonstration of the inequalities (iii) and (iv) of Theorem 4.2.] □

5. An example of an estimation of the real roots of a polynomial

Let us consider the polynomial

$$f(x) = x^6 - 12x^4 - 2x^3 + 37x^2 + 10x - 10 .$$

The successive derivatives of f are

$$\begin{aligned} f'(x) &= 2(3x^5 - 24x^3 - 3x^2 + 37x + 5), \\ f''(x) &= 2(15x^4 - 72x^2 - 6x + 37), \\ f^{(3)}(x) &= 12(10x^3 - 24x - 1), \\ f^{(4)}(x) &= 72(5x^2 - 4), \\ f^{(5)}(x) &= 720x, \quad \text{and} \quad f^{(6)}(x) = 720 . \end{aligned}$$

Let us apply the various rules that we have seen to polynomial f.

Rule of Newton : We must find a real number c such that $f^{(i)}(c)$ is positive for $i = 1, 2, \ldots, 6$. It is easy to verify that we have $f^{(i)}(c) \geq 0$ for $i = 2, \ldots, 6$ and $c \geq \sqrt{7}$. The inequality $f(\sqrt{8}) > 0$ shows that we can take the value $L = \sqrt{8} = 2.828\ldots$ as an upper bound of the roots.

Rule of Lagrange and MacLaurin : We find at once the upper bound $1 + \sqrt{12} = 4.464\ldots.$

Rule of Cauchy : With the notations of Proposition 5.5, we have here $k = 3$ and we find the bound

$$\max\left\{(3 \times 12)^{1/2}, (3 \times 2)^{1/3}, (3 \times 10)^{1/6}\right\} = 6.$$

Rule of Laguerre (see Exercise 5.12) : It gives the bound $\sqrt{13} = 3.605\ldots$

4. The number of zeros of a polynomial in a real interval

1. The rule of Sturm

Before stating the theorem of Sturm, we need two new definitions.

Let f be a polynomial and let a and b be two real numbers, where $a < b$. We say that a sequence of polynomials $f_0 = f, f_1, \ldots, f_s$ is a *sequence of Sturm* for f on the interval $[a, b]$, when the following conditions are satisfied :

(i) $f(a) \cdot f(b) \neq 0$.

(ii) The polynomial function f_s has no zero on the closed real interval $[a, b]$.

(iii) If c is a real number such that $a < c < b$ and $f_j(c) = 0$ for an index j, with $0 < j < s$, then we have $f_{j-1}(c)f_{j+1}(c) < 0$.

(iv) If c is a real number such that $a < c < b$ and $f(c) = 0$, then the product polynomial $f(x) \cdot f_1(x)$ has the same sign as $x - c$ in the neighborhood of the point c.

To such a sequence and to a point x of the interval $]a, b[$, we associate the number of variations of sign of this sequence at the point x; this number is written $\mathrm{V}(f_0, \ldots, f_s\,;\, x)$, or in a simpler way $\mathrm{V}(x)$, which means that $\mathrm{V}(x)$ is defined by the formula

$$\mathrm{V}(x) = \mathrm{Card}\left\{(i, j)\,;\, f_i(x)f_j(x) < 0 \quad \text{and} \quad f_k(x) = 0 \text{ if } \ i < k < j\right\},$$

where the indices i and j satisfy $0 \leq i < j \leq n$.

We can, now, state the theorem of Sturm.

THEOREM 5.3 (Sturm). — *Let f be a polynomial with real coefficients and let a and b be two real numbers a and b, where $a < b$. If f_0, f_1, $\ldots$, f_s is a sequence of Sturm for f on the real interval $\left[a, b\right]$. Then the number of distinct zeros of f on this interval is equal to $\mathrm{V}(a) - \mathrm{V}(b)$.*

Proof

Let us call I the interval $\left[a, b\right]$. We first notice that $\mathrm{V}(x)$ is constant on any interval contained in I and on which none of the functions f_i becomes zero. Since the functions f_i only have a finite number of zeros, we only have to show that the quantity $\mathrm{V}(x)$ has the following properties :

(a) $\mathrm{V}(x)$ decreases by 1 when x passes through a root c of f, when the point c belongs to the interval I,

(b) $\mathrm{V}(x)$ does not change when x crosses a point c of I such that $f(c)$ is nonzero and $f_i(c) = 0$ for at least one index $i \geq 1$.

Suppose, at first, that c is a root of f. By (iii), we have $f_1(c) \neq 0$. Whereas, by (iv), for $h \neq 0$ small enough, we have

$$\operatorname{sign}\left\{f(c+h) \cdot f_1(c+h)\right\} = \operatorname{sign}(h),$$

which shows that there is a change of sign at least at the beginning of the sequence of $f_i(x)$ (in fact between the indices 0 and 1) when x varies from the value $c - \varepsilon$ to $c + \varepsilon$, for any small enough positive ε.

Suppose, now, that c is a root of f_i, with $i \geq 1$, and $c \in I$. By (ii), we have $i < s$. By (iii), we have $f_{i-1}(c) \cdot f_{i+1}(c) < 0$. Hence, there is no variation in the number of changes of sign of the $f_j(x)$, for $i-1 \leq j \leq i+1$, when the point x crosses c [in fact, we have $f_{i-1}(x) \cdot f_{i+1}(x) < 0$ when x is close to c].

The information above implies properties (a) and (b). Thus, the theorem is demonstrated. $\square$

The following result gives a way to build a sequence of Sturm associated to a polynomial and hence to apply the previous theorem.

PROPOSITION 5.6 (Sturm). — *Let f be a polynomial with real coefficients and let a and b be any two real numbers, with $a < b$. Consider the sequence $f_0,\ f_1,\ \ldots,\ f_{s+1}$ of polynomials defined by the conditions*

$$f_0 = f, \quad f_1 = f', \quad f_2 = -f \bmod f', \ \ldots,$$
$$f_{i+1} = -f_{i-1} \bmod f_i, \quad \ldots, \quad f_{s+1} = 0,$$

that is, up to signs, the sequence we get by computing the g.c.d. of the polynomials f and f' with the algorithm of Euclid.

Then the sequence of polynomials $g_i = f_i/f_s,\ 0 \le i \le s,$ is a sequence of Sturm for g_0 on the interval $[a, b]$, as long as we have $g_0(a) \cdot g_0(b) \ne 0$.

Proof

We have to verify the properties (i) to (iv) above.

(i) : True when $g_0(a)\, g_0(b) \ne 0$.

(ii) : Direct consequence of the fact that $g_s = 1$.

(iii) : By construction, we have relations like

$$g_{j-1}(X) = q_j(X) g_j(X) - g_{j+1}(X) \quad \text{for} \quad 0 < j < s.$$

Hence $g_{j-1}(c)\, g_{j+1}(c) < 0$ when $g_j(c) = 0$ (notice that c can be neither a root of g_{j-1} nor of g_{j+1}, otherwise c would be a root of $g_s = 1$).

(iv) : We have $g_1(c) \ne 0$, and thanks to Corollary 5.2 (after the theorem of Rolle), we conclude. ▯

COROLLARY. — *Let f be a polynomial with real coefficients and let a and b be any two real numbers, $a < b$. Consider the sequence $f_0,\ f_1,\ \ldots,\ f_{s+1}$ of real polynomials defined by $f_0 = f$ and*

$$f_1 = f',\ f_2 = -f \bmod f',\ \ldots,\ f_{i+1} = -f_{i-1} \bmod f_i,\ \ldots,\ f_{s+1} = 0.$$

Then, if $f(a)$ and $f(b)$ are nonzero, the number of distinct zeros of f in the interval $[a, b]$ is equal to the difference

$$\mathrm{V}(f_0, f_1, \ldots, f_s\,;\, a) - \mathrm{V}(f_0, f_1, \ldots, f_s\,;\, b).$$

Proof

The proof is an easy exercise. ▯

2. The theorem of Budan-Fourier

This result is less precise, but easier to use than the theorem of Sturm. We shall see in an example that it is sometimes powerful enough to separate the real roots of a polynomial.

THEOREM 5.4 (Budan-Fourier). — *Let f be a nonconstant polynomial with real coefficients and let a and b, with $a < b$, be two real numbers such that $f(a) \cdot f(b) \neq 0$. Designate by $\mathrm{v}(x)$ the number of changes of signs in the sequence $f(x)$, $f'(x)$, $f''(x)$, $\ldots$, $f^{(n)}(x)$. Then the number of zeros of f in the interval $[a, b]$, counted with their orders of multiplicity, is equal to*

$$\mathrm{v}(a) - \mathrm{v}(b) - 2m, \quad \textit{for some} \;\; m \in \mathbb{N}.$$

Proof

Let us call I the interval $[a, b]$. We first notice that the function $\mathrm{v}(x)$ is constant on any subinterval of I on which none of the functions $f^{(i)}$ becomes zero. Since the functions $f^{(i)}$ only have finitely many zeros, we only have to demonstrate that the quantity $\mathrm{v}(x)$ satisfies the two following properties :

(i) $\mathrm{v}(x)$ decreases of $k + 2m'$, for some $m' \in \mathbb{N}$, when x crosses a root c of order k of f, for a point c in I,

(ii) $\mathrm{v}(x)$ decreases of $2m''$, $m'' \in \mathbb{N}$, when x crosses a point c of I such that $f(c)$ is nonzero and $f^{(i)}(c) = 0$ for at least one index $i > 1$.

First, suppose that c is a root of order k of f. Then $f^{(k)}(c)$ is nonzero and, by Taylor's formula, we have the estimate

$$f^{(i)}(x) \approx \frac{k(k-1)\cdots(k-i+1)}{k!}\,(x-c)^{k-i}\, f^{(k)}(c), \quad \text{for} \;\; 0 \leq i \leq k,$$

for x close to c, $x \neq c$. Which shows that there are k changes of sign at least at the beginning of the sequence of the $f^{(i)}(x)$ (in fact between the indices 0 and k) when the point x varies from the value $c - \varepsilon$ to $c + \varepsilon$, for $\varepsilon > 0$ small enough.

Suppose, now that c is a root of order k of $f^{(i)}$ with $i \geq 1$, for some point c of the interval I, and that we have $f^{(i-1)}(c) \neq 0$. Again, we apply the formula of Taylor at the point c, but this time to function $f^{(i)}$. Let us consider the two cases where k is even and where k is odd.

When k is even, $f^{(i)}(x)$ does not change sign in the neighborhood of c (for $x \neq c$). We notice that the sequence $f^{(i)}(x), \ldots, f^{(i+k)}(x)$ has k changes of sign at least when the point x varies from the value $c-\varepsilon$ to $c+\varepsilon$, for $\varepsilon > 0$ small enough.

When k is odd, $f^{(i)}(x)$ changes sign around the point c (for $x \neq c$). We see that the sequence $f^{(i-1)}(x), f^{(i)}(x), \ldots, f^{(i+k)}(x)$ looses $k-1$, or $k+1$, changes sign when the point x varies from the value $c-\varepsilon$ to $c+\varepsilon$, for $\varepsilon > 0$ small enough, in function of the sign of $f^{(i-1)}(c)$ (note that this sign does not change in a small neighborhood of the point c).

This information that we have just found imply properties (i) and (ii). Hence we have proved the theorem. □

3. The rule of Descartes

THEOREM 5.5 (Descartes). — *Let $f = a_n X^n + a_{n-1} X^{n-1} + \cdots + a_0$ be a polynomial with real coefficients, where a_n and a_0 are nonzero. Let v be the number of changes of signs in the sequence $(a_0, \ldots, a_n)$ of its coefficients and let r be the number of its real positive roots, counted with their orders of multiplicity, then there exists some integer nonnegative integer m such that*

$$r = v - 2m .$$

Proof

Let M be a bound of the positive zeros of f and let L be a real number, with $L \geq M$. At the point zero, we have $f^{(i)}(0) = i!\, a_i$. Whereas, for L big enough, all the $f^{(i)}(L)$ are of the sign of a_n. The conclusion follows from the theorem of Budan-Fourier applied in the interval $[0, L]$. □

COMPLEMENT. — *With the notations of the theorem, when v is equal to 0 or to 1, the number of positive roots of the polynomial is equal to v.*

Proof

This is an easy consequence of the two following properties : $r = v - 2m$ and $r \geq 0$. Notice that the case $v = 1$ has already been demonstrated twice : Proposition 5.3 above and lemma 4.1. □

Example

Consider the polynomial $f(x) = x^6 - x^4 + 2x^2 - 3x - 1$. For this polynomial, we have $v = 3$. The number of positive roots of f is equal to 1 or 3. It is obvious that $f(x)$ takes only negative values on the closed interval $0 \leq x \leq 1$. Let us write $x = 1 + y$; then, for example, by applying Taylor's formula, we get the formula

$$f(1+y) = -2 + 3y + 11y^2 + 16y^3 + 14y^4 + 6y^5 + y^6,$$

which shows that f has only one real positive root.

COROLLARY. — *A polynomial with real coefficients with m nonzero coefficients has at most $m-1$ positive real roots and also at most $m-1$ strictly negative roots (we have the same result when we consider the orders of multiplicity).*

Proof

Indeed, a sequence that has exactly m nonzero terms does not have more than $m-1$ changes of sign. □

4. Case of the polynomials whose all roots are real

We keep the notations of Theorem 5.5. Let v' be the number of sign variations of the list of coefficients of the polynomial $f(-x)$, that is the number of variations of sign of the list $a_0, -a_1, \ldots, (-1)^n a_n$.

First, notice that if none of the a_i is zero then $v + v' = n$. Indeed, for each of the two lists $(a_0, a_1, \ldots, a_n)$ and $(a_0, -a_1, \ldots, (-1)^n a_n)$, there are n comparisons to do to detect the changes of sign and, when there is one change of sign at some position in the first list, there is no change of sign at this position in the second list, and *vice-versa*.

We show that always $v + v' \leq n$. In order not to have heavy notations, we argue by induction on the number k of zero coefficients of the sequence $a_0, a_1, \ldots, a_n$. The case where this number is zero has just been studied.

Suppose that $1 \leq k \leq n$ and that the result is true for $k-1$. Let i be the smallest index such that a_i is equal to zero. Then, consider the list $a_0, a_1, \ldots, a_{i-1}, a_{i-1}, a_{i+1}, \ldots, a_n$: it is obvious that, on one hand, this list also has exactly v changes of signs and that, on the other hand, the number v'' of changes of signs of the sequence a_0, $-a_1$, $\ldots$, $(-1)^{i-1}a_{i-1}$, $(-1)^i a_{i-1}$, $(-1)^{i+1}a_{i+1}$, $\ldots$, $(-1)^n a_n$ is at least equal to v'. By induction, we have $v+v'' \leq n$. Hence the inequality is obvious.

The following result can be deduced from what we have just seen and from the rule of Descartes.

PROPOSITION 5.7. — *Let $f(x)$ be a polynomial with real coefficients, we call v the number of changes of signs of the list of its coefficients and we call v' the number of changes of signs of the list of the coefficients of the polynomial $f(-x)$. Then, if all the roots of f are real and nonzero, the number r of positive roots of f and the number r' of its negative roots satisfy*

$$r = v \quad \textit{and} \quad r' = v' .$$

Proof

By applying the rule of Descartes to the polynomials $f(x)$ and $f(-x)$, we get the inequalities

$$r \leq v \quad \text{and} \quad r' \leq v' .$$

Moreover, in the demonstration above, we have seen that

$$v + v' \leq \deg(f) .$$

Since all the roots of f are real and nonzero, we also have

$$r + r' = \deg(f) .$$

It follows that the previous inequalities are, in fact, equalities. This ends the proof. □

5. A detailed example

Consider the polynomial

$$f(x) = x^5 - 5x^4 + 15x^3 - x^2 + 3x - 7 .$$

We want to determine the number of real roots of f and to separate them. Therefore, first compute a Sturm sequence of this polynomial. We successively get

$$\begin{aligned} f_0 = f(x) &= x^5 - 5x^4 + 15x^3 - x^2 + 3x - 7, \\ f_1(x) = f'(x) &= 5x^4 - 20x^3 + 45x^2 - 2x + 3, \\ 5f_0 &= (x-1)\, f_1 + (10x^3 + 42x^2 + 10x - 32). \end{aligned}$$

Hence, we can take

$$f_2 = -5x^3 - 21x^2 - 5x + 16,$$

then

$$5f_1 = (-5x + 41)\, f_2 + (1061x^2 + 275x - 641),$$

and thus,

$$f_3 = -1061x^2 - 275x + 641.$$

Notice that all the roots of the polynomial f_3 belong to the closed real interval $\left[-1, +1\right]$.

First, consider the interval $\left[1, +\infty\right[$. For $x \geq 1$, we have

$$f(x) = x^5 - 5x^4 + 15x^3 - x^2 + 3x - 7 > x^5 - 5x^4,$$

hence f has no real root ≥ 5. Moreover, we have

$$\mathrm{V}\,(f_0, f_1, f_2, f_3\,;\, 1) = \mathrm{V}\,(6, 29, -15, -695) = 1,$$

whereas, for b large enough, the sequence of the signs of $f_i(b)$, $i = 0, 1, 2, 3$, is $(+, +, -, -)$ and hence

$$\mathrm{V}\,(f_0, f_1, f_2, f_3\,;\, b) = 1.$$

Which shows that the polynomial f has no root in the interval $\left[1, +\infty\right[$.

Other proof : we have

$$\begin{aligned} f(x) &= x^5 - 5x^4 + 15x^3 - x^2 + 3x - 7\,, \\ f'(x) &= 5x^4 - 20x^3 + 45x^2 - 2x + 3\,, \\ f''(x) &= 2(10x^3 - 30x^2 + 45x - 1)\,, \\ f^{(3)}(x) &= 30(2x^2 - 4x + 3)\,, \\ f^{(4)}(x) &= 120(x-1), \quad f^{(5)}(x) = 120\,, \end{aligned}$$

and hence

$$f^{(i)}(1) \geq 0 \quad \text{for} \quad i = 0, 1, \ldots, 5\,,$$

which implies (thanks to the rule of Newton) that the polynomial has no root in the interval $\left[1, +\infty\right[$.

Consider now the interval $\left[0, +1\right]$. For x in this interval , we have

$$f'(x) = (45x^2 - 20x^3 + 5x^4) + (3 - 2x) > 0\,;$$

hence f is strictly increasing on this interval, and, since $f(0) = -7$ and $f(1) = 6$, the polynomial f has exactly one root in the interval $\left]0, +1\right[$. Notice that the rule of Budan-Fourier shows that f has one or three roots in the interval $\left]0, +1\right[$.

Consider now the infinite interval $\left]-\infty, 0\right]$. All the coefficients of the polynomial $f(-x)$ are negative; hence f has no zero in this interval (trivial case of the rule of Descartes).

We can easily give a precise approximation of the real root ξ of f. For example, we notice that

$$f\Big(\frac{1}{\sqrt{2}}\Big) = \frac{1}{4\sqrt{2}} - \frac{5}{4} + \frac{15}{2\sqrt{2}} - \frac{1}{2} + \frac{3}{\sqrt{2}} - 7 = \frac{43}{4\sqrt{2}} - \frac{35}{4} > 0\,.$$

which shows that ξ is lying between $1/\sqrt{2}$ and 1.

Since $f'(1) = 31$, the method of Newton, applied with $x_0 = 1$ as first point, gives the approximation

$$x_1 = x_0 - f(x_0)/f'(x_0) = 1 - 6/31 = 0.806\ldots$$

By applying twice the method of Newton with the value 0.8 as first point, we get the approximate value $\xi \approx 0.766$.

The measure of f satisfies

$$\mathrm{M}(f) < (1 + 5^2 + 15^2 + 1 + 3^2 + 7^2)^{1/2} = 310^{1/2} < 17.61;$$

hence the complex roots of f have a maximal modulus < 4.2. Indeed, if z is an imaginary root of f, then its complex conjugate is also a root of f and we have

$$\mathrm{M}(f) \geq |z|^2,$$

and, hence, here

$$|z| < \sqrt{17.61} < 4.2.$$

6. *Vincent's theorem*

THEOREM 5.6 (Vincent). — *Let f be a polynomial with real coefficients without multiple roots and let $a_1, \ldots, a_k, \ldots$ be an arbitrary sequence of positive rational integers. By transforming the polynomial $f(x)$ with the successive changes of variables*

$$x_i = a_i + \frac{1}{x_{i+1}}, \quad i = 0, 1, \ldots \quad (\textit{where } x_0 = x),$$

we get, after a finite number of steps, a polynomial whose list of coefficients has at most only one change of sign.

Proof

Consider the continued fraction defined by the sequence $a_1, \ldots, a_k, \ldots$ Let a be the limit of this continued fraction; there are two cases :

(i) a is not a root of f,

(ii) we have $f(a) = 0$.

Let us call p_k/q_k the kth principal convergent of the development in continued fraction|| of a. After m changes of variables, as in the statement of the theorem, if the new variable is written y, we have

$$x = \frac{p_m y + p_{m-1}}{q_m y + q_{m-1}},$$

which is equivalent to

$$y = \frac{p_{m-1} - q_{m-1}x}{p_m - q_m x}.$$

We know that, for any positive integer k, we have

$$\left|\alpha - \frac{p_k}{q_k}\right| \leq \frac{1}{q_{k+1}\, q_k} < \frac{1}{{q_k}^2},$$

and thus, the estimate

$$y = -\frac{q_{m-1}(\alpha - x) + \mathrm{O}(1/q_m)}{q_m(\alpha - x) + \mathrm{O}(1/q_m)}.$$

Let us make again a change of variable, namely

$$y = \frac{q_{m-1}}{q_m u}.$$

Let $F(y)$ be the polynomial obtained from the polynomial f by the change of variable $x \longmapsto y$ and let $G(u)$ be the polynomial obtained from F by the change of variable $y \longmapsto u$. Then, the polynomials F and G have the same number of changes of signs in the list of their coefficients.

From the previous estimate of y, it follows at once that the roots of G transformed from the roots of f other than α are arbitrarily close to -1 when m is big enough. Consequently, in case (i), the polynomial G has coefficients which are all of the same sign and Vincent's theorem is demonstrated.

|| For a reference on continued fractions, see Hardy and Wright, *loc. cit.*, Chap. 6.

Consider now the second case. Let ω be the root of G associated to α, we have

$$\omega = -\frac{\dfrac{p_{m-1}}{q_{m-1}} - \alpha}{\dfrac{p_m}{q_m} - \alpha},$$

which implies $\omega > 1$. Up to a multiplicative constant, the polynomial G is equal to

$$G(u) = (u - \omega)\,(u + 1 + \varepsilon_1) \cdots (u + 1 + \varepsilon_{n-1}),$$

where the ε_i are arbitrarily small for m big enough and where n designates the degree of f.

By Lemma 5.1 below, the sequence of the quotients of consecutive coefficients of the polynomial $(u+1)^{n-1}$ is decreasing; indeed, an easy direct computation shows that it is strictly decreasing. The same property holds, at least when m is big enough, for the polynomial

$$(u + 1 + \varepsilon_1) \cdots (u + 1 + \varepsilon_{n-1}).$$

Finally, a new application of Lemma 5.1 shows that the list of the coefficients of the polynomial G has exactly one change of sign and the theorem is demonstrated. □

LEMMA 5.1. — *Let*

$$f(x) = a_0 x^n + a_1 x^{n-1} + \cdots + a_n$$

be a polynomial with real positive coefficients. Then, the two following properties hold :

1°) If all the roots of f are real, then the sequence of the quotients a_{i+1}/a_i is decreasing.

2°) If the sequence of the quotients of consecutive coefficients a_{i+1}/a_i is decreasing, then for any positive real number ξ, the list of the coefficients of the product polynomial $(x - \xi)\,f(x)$ has exactly one change of sign.

Proof

First suppose that all the roots of f are real; since the coefficients of the polynomial f are all positive, its roots are necessarily negative. Let ξ be any positive real number. Since the product polynomial $(x-\xi)f(x)$ has only real roots, with only one which is positive, Proposition 5.7 shows that the list of its coefficients has exactly one change of sign. In other words, for any positive number ξ, there exists one index j, such that

$$i < j \Longrightarrow a_{i+1} - \xi a_i \geq 0 \quad \text{and} \quad i > j \Longrightarrow a_{i+1} - \xi a_i \leq 0,$$

where we have written $a_{-1} = 0$, which shows, indeed, that the sequence of the quotients a_{i+1}/a_i is decreasing. This proves the first assertion.

Suppose now that the sequence of the quotients a_{i+1}/a_i is decreasing. Then, for any real number $\xi > 0$, there exists an index j such that

$$i < j \Longrightarrow a_{i+1} - \xi a_i \geq 0 \quad \text{and} \quad i > j \Longrightarrow a_{i+1} - \xi a_i \leq 0,$$

where again $a_{-1} = 0$. Which shows, indeed, that the sequence of the coefficients of the polynomial $(x-\xi)f(x)$ has exactly one change of sign. Hence, we have the second property. □

Vincent proposes the following method to separate the real roots of a polynomial f with real coefficients and without multiple roots.

1°) Determine two positive integers L and L' such that all the real roots of f are between $-L'$ and L. In order to simplify the notations, we just consider the separation of the positive roots. Studying the polynomial $f(-x)$, we can, however, just look for positive real roots.

2°) Let, then a and a' be two rational integers with $0 \leq a < a' \leq L$. We know, by the theorem of Budan-Fourier, that the polynomial f can have a root in the interval $\left[a, a'\right]$ only if we have $\mathrm{v}(a) > \mathrm{v}(a')$ where — let us remind it — the notation $\mathrm{v}(x)$ designates the number of changes of signs in the sequence $f(x)$, $f'(x)$, $f'(x)$, …, $f^{(n)}(x)$, and where n is the degree of the polynomial f. To simplify, say that $\mathrm{v}(x)$ is the variation of f at the point x. When it is useful to mention which is the considered polynomial, this variation is written $\mathrm{v}(f\,;\,x)$.

If we suppose that $a' = a + 1$, and that we have $\mathrm{v}(a) > \mathrm{v}(a')$, we write

$$x = a + \frac{1}{x_1} \cdot$$

In this way, we get a polynomial f_1 in x_1 such that the roots of f belonging to the interval $[a, a+1]$ correspond to the roots > 1 of f_1. If r is the number of these roots, then we have

$$r \leq \mathrm{v}(f_1\,;\,1).$$

Then, we try to separate the roots > 1 of the polynomial f_1, applying the same process ... and so on.

This procedure produces the beginning of the development in continued fractions of each of the real roots of f. And, thanks to Vincent's theorem, we can suppose that all these computations have been done until, for each of the real roots of the polynomial f, we have got a transformed polynomial that has exactly a single variation of sign.

3°) For some real root of f, let us call g the last computed transformed polynomial of f and y the variable associated to g. We suppose that the variation of g is equal to 1 and that we have found an approximate value γ such that the unique positive root of the polynomial g is equal to $\gamma + h$, with $0 < h \leq \varepsilon < 1$. By the theorem of Budan-Fourier, we have $\mathrm{v}(g\,;\,\gamma) = 1$. We suppose also, which is always possible, that the leading coefficient of g is positive. Finally, we add the condition

$$g'(\gamma) > 0,$$

which is satisfied for h small enough, since we have

$$\xi \approx \gamma - g(\gamma)/g'(\gamma), \quad \text{where} \quad \gamma < \xi,$$

and hence $g(\gamma) < 0$.

The conditions $\mathrm{v}(g\,;\,\gamma) = 1$ and $g'(\gamma) > 0$ imply

$$g^{(i)}(\gamma) \geq 0, \quad \text{for } i = 1, 2, \ldots, n,$$

and Taylor's development of $g(\gamma+h)$ shows that we have

$$\xi < \gamma - g(\gamma)/g'(\gamma).$$

Finally, a classical error computation leads to the inequality

$$\xi > \gamma - \frac{g(\gamma)}{g'(\gamma)} - \frac{1}{2}\frac{g''(\gamma+\varepsilon)}{g'(\gamma)}\,\varepsilon^2 .$$

This allows us to find rapidly a good approximate value for the number ξ and, consequently, thanks to the computation of x in terms of y done during the proof of Vincent's theorem, to find a good approximate value for the considered root α of f.

Let us take an example from the original article of Vincent, which was studied earlier by Lagrange, that is the polynomial

$$f(x) = x^3 - 7x + 7.$$

Step 1. Determine an upper bound L for the positive roots of f. Notice that

$$x \geq 2 \implies f'(x) = 3x^2 - 7 > 0,$$

which proves that

$$x \geq 2 \implies f(x) \geq f(2) = 1,$$

and hence the value $L = 2$ fits.

Step 2. Compute the $f^{(i)}(x)$. This gives the following table

x	$f(x)$	$f'(x)$	$\frac{1}{2}f''(x)$	$\frac{1}{6}f^{(3)}(x)$
0	7	−7	0	1
1	1	−4	3	1
2	1	5	6	1

The table shows that the positive real roots belong to the interval $]1,2[$ (this gives also another proof that the value $L = 2$ is a correct upper bound for the real roots of f) and that this interval contains zero or two roots.

Hence, we now transform the equation by the new change of variable $x = 1 + {x_1}^{-1}$, which gives the polynomial

$$f_1 = {x_1}^3 - 4{x_1}^2 + 3x_1 + 1,$$

which can be read directly on the previous table. We compute a table similar to the previous one for the polynomial f_1 :

x_1	$f_1(x_1)$	$f_1'(x_1)$	$\frac{1}{2}f_1''(x_1)$	$\frac{1}{6}f_1^{(3)}(x_1)$
1	1	-2	-1	1
2	-1	-1	2	1
3	1	6	5	1

This table shows that f_1 has a root between 1 and 2 and another one between 2 and 3. Which leads us to make the first change of variable $x_1 = 1 + x_2^{-1}$. This gives the polynomial

$$f_2 = {x_2}^3 - 2{x_2}^2 - x_2 + 1,$$

and the new table

x_2	$f_2(x_2)$	$f_2'(x_2)$	$\frac{1}{2}f_2''(x_2)$	$\frac{1}{6}f_2^{(3)}(x_2)$
1	-1	-2	1	1
2	-1	3	4	1
3	7	14	7	1

which shows that f_2 has exactly one root in the real interval $]2, 3[$.

The change of variable $x_2 = 2 + y^{-1}$ leads to the polynomial

$$g(y) = y^3 - 3y^2 - 4y - 1,$$

whose coefficients only have one sign variation, between the first and the second term, as we wanted. Hence the polynomial g has one, and only one positive root β, which is situated between 4 and 5 [since we have $g(4) < 0$ but $g(5) > 0$].

We also have to make the change of variable $x_1 = 2 + z^{-1}$, which leads to the polynomial

$$h(z) = z^3 + z^2 - 2z - 1\,,$$

which can also be read on the second table. The coefficients of this polynomial only have one sign variation. Hence the polynomial h has one, and only one, positive root β', which is situated between 1 and 2 [since we have $h(1) = -1$ and $h(2) = 7$].

Now, let us estimate β and β' with a little more precision. A first application of the Newton's formula gives the values

$$\beta \approx 4 + 1/20 = 4.05$$

and

$$\beta' \approx 1 + 1/3 \approx 1.33\,,$$

the precision for the first value seems to be better than that of the second value.

Let α and α' be the roots of f respectively associated to β and to β'. To have a sharper precision for the root α', it is better to transform again the polynomial f. Since the positive root of h is lying between 1 and 2, we take t as new variable defined by $z = 1 + t^{-1}$. Which gives, the polynomial

$$k(t) = t^3 - 3t^2 - 4t - 1\,,$$

which is the polynomial $g(t)$ that we met already above.

Let us come back to polynomial f. We have the formulas

$$\alpha = 1 + \cfrac{1}{1 + \cfrac{1}{2 + \beta^{-1}}}$$

and

$$\alpha' = 1 + \frac{1}{2 + \beta'^{-1}} = 1 + \cfrac{1}{2 + \cfrac{1}{1 + \beta^{-1}}}\,.$$

Hence the inequalities

$$\frac{27}{26} < \alpha < \frac{22}{13}, \qquad \frac{23}{17} < \alpha' < \frac{19}{14},$$

and the approximate values

$$\alpha \approx 1.6920, \quad \alpha' \approx 1.3569.$$

Other strategies are possible instead of Vincent's :

- Uspensky* proposes to use only the transformations $x = 1 + y$ and $x = (1+z)^{-1}$, which allows to separate the zeros of f between 0 and 1 from those > 1. The same procedure is then applied to the transformed polynomials, which, at first sight, seems to be less efficient than the procedure of Vincent.

- Collins and Akritas† propose a technique of dichotomy to separate the roots, mainly between these of the interval $[0, L/2]$ and those of the interval $[L/2, L]$, and so on ..., saves computations. See their article for a precise analysis.

5. Equations whose roots have a negative real part

This question is related to the stability of mechanical systems. It was first presented by Maxwell, first treated by Routh, and completely solved by Hurwitz, whose result is the following.

THEOREM 5.7. — *Consider a polynomial with real coefficients*

$$f(x) = p_0 + p_1 x + \cdots + p_n x^n, \quad p_0 > 0, \ p_n \neq 0.$$

Then all the roots of f have a negative real part if, and only if, we have the following inequalities

$$D_i = \begin{vmatrix} p_1 & p_0 & 0 & \cdots & 0 \\ p_3 & p_2 & p_1 & \cdots & 0 \\ \cdots & \cdots & \cdots & \cdots & \cdots \\ p_{2i-1} & p_{2i-2} & p_{2i-3} & \cdots & p_i \end{vmatrix} > 0, \qquad i = 1, 2, \ldots, n,$$

where we write $p_j = 0$ if $j < 0$ or $j > n$.

---

* J.V. Uspensky, *Theory of equations*, McGraw-Hill, New York, 1948.

† Cf. G. Collins and R. Loos, in *Computer Algebra*, ed. by B. Buchberger, G. Collins, and R. Loos, Springer-Verlag, Wien, 1982.

Proof

We follow the demonstration given by I. Schur. Argue by induction on the degree n of the polynomial f.

- If n is equal to 1 then $f(x) = p_0 + p_1 x$ and $D_1 = p_1$; hence the theorem is true for $n = 1$.

- The following lemma will be useful to treat the general case.

LEMME 5.2. — *If f is a polynomial with real coefficients that only has roots which real parts are negative, then, for any complex number z we have*

$$\operatorname{sign}\{|f(z)| - |f(-z)|\} = \operatorname{sign}\{\operatorname{Re}(z)\}.$$

Proof

It is enough to study only the case where the real part of z is positive. Write $z = a + ib$, with a and b real and $a \geq 0$. Let α be any complex number, $\alpha = \beta + i\gamma$, with β and γ real and $\beta < 0$. Then, we have the relation

$$|(z-\alpha)(z-\bar{\alpha})|^2 = (a-\beta)^4 + 2(a-\beta)^2(b^2+\gamma^2) + (b^2-\gamma^2)^2.$$

By changing z into $-z$, the right-hand side remains invariant if a is zero and decreases when a is positive. Now, the lemma is obvious by applying this remark for each root of the polynomial f. □

Let us come back to the proof of theorem 5.7. We shall construct a polynomial ψ of degree $< n$ whose roots all have a negative real part if, and only if, the same property holds for the polynomial f. We can then argue by induction on the degree of the polynomial. Define the polynomial ψ by the formula

$$\psi = p_0 p_1 + \omega(p_1 p_2 - p_0 p_3)x + p_0 p_3 x^2 + \omega(p_1 p_4 - p_0 p_5)x^2 + \cdots;$$

hence, we have the relation

$$2x\psi = A(x)\, f(x) - B(x)\, f(-x)$$

with

$$A(x) = \{p_0(1-\omega/x) + p_1\omega\} \quad \text{and} \quad B(x) = \{p_0(1-\omega/x) - p_1\omega\}.$$

We verify that ψ is a polynomial of degree $n-1$, and we get the implication

$$\left(p_0 > 0,\ p_1 > 0,\ \omega > 0,\ z \in \mathbb{C}^*,\ \operatorname{Re}(\omega/z) < 1\right) \Longrightarrow |A(z)| > |B(z)|.$$

If $\omega \in \left]0, 1\right[$ is small enough, for any root ξ of ψ, we have $\operatorname{Re}(\omega/\xi) < 1$ because of Lemma 5.2 and of the implication above. Hence, we have the following consequence

$$\psi(\xi) = 0 \Longrightarrow A(\xi)f(\xi) = B(\xi)f(-\xi), \quad \text{with} \quad |A(\xi)| > |B(\xi)|.$$

This implies the inequality $\operatorname{Re}(\xi) < 0$, since all the roots of f have a negative real part.

On the other hand, we can verify that

$$2p_0p_1\omega f(x) = xA(-x)\psi(x) - xB(x)\psi(-x).$$

Let α be a root of f whose real part is non negative; then we have

$$A(-\alpha)\psi(\alpha) = B(\alpha)\psi(-\alpha),$$

and, if $\alpha^{-1} = \beta + i\gamma$, with $\beta \geq 0$,

$$\begin{aligned} A(-\alpha) &= p_0(1 + \beta\omega + i\gamma\omega) + p_1\omega, \\ B(\alpha) &= p_0(1 - \beta\omega - i\gamma\omega) - p_1\omega, \end{aligned}$$

so that

$$|A(-\alpha)|^2 - |B(\alpha)|^2 = 4\omega p_0^2\beta + 4p_0p_1\omega > 0.$$

This last inequality implies that

$$|A(-\alpha)| > |B(\alpha)|$$

and finally, we obtain the inequality

$$|\psi(\alpha)| > |\psi(-\alpha)|.$$

By the lemma, this last inequality is impossible when all the roots ξ of ψ have a negative real part. Hence, we can apply the induction hypothesis to the polynomial ψ.

Finally, we verify that the determinants associated with ψ are

$$\Delta_1 = \omega(p_1p_2 - p_0p_3) = \omega D_2\,,$$
$$\Delta_2 = \omega p_0\big((p_1p2 - p_0p_3)p_3 - (p_1p_4 - p_0p_5)p_1\big) = \omega p_0 D_3\,,$$

and, more generally, that the determinant Δ_k is given by a formula like

$$\Delta_k = \lambda_k D_{k+1}\,,$$

where the factor λ_k is positive. □

Exercises

1. Demonstrate the rule of Descartes. Argue by induction on the degree n of the polynomial f and consider the derivative f' of f.

[If $n \leq 1$, then, the result is obvious. If $n > 1$ and if the polynomial f is given by $f = a_n X^n + a_{n-1} X^{n_1} + \cdots + a_0$, with $a_0\, a_n \neq 0$, then, we can write

$$f' = n a_n X^{n-1} + (n-1) a_{n-1} X^{n-2} + \cdots + q a_q X^{q-1},$$

with $a_q \neq 0$. By induction hypothesis, the number r' of positive zeros of f' is equal to $v' - 2m'$, for some integer $m' \in \mathbb{N}$.

If f' has no positive zero, then, on one hand v' is even and on the other hand $r = 0$ or 1 whether the product $a_n\, a_0$ is positive or negative ...

If f' has at least one positive zero, use Rolle's theorem on the closed real interval $\big[0, c\big]$, where c is the smallest positive zero of f' ...]

2. This exercise aims, first of all, to show the following lemma, due to Segner :

> If we call $\mathrm{V}(P)$ the number of changes of sign of the sequence of the coefficients of a real monic polynomial P, then, for any positive real number c, the number of changes of sign of the sequence of the coefficients of the polynomial $Q(X) = (X-c)P(X)$ is given by the formula $V(P) + 1 + 2m$, where m is some nonnegative rational integer.

[Put $P = X^n + a_1 X^{n-1} + \cdots + a_n$. Let $V = \mathrm{V}(P)$ and let $i_1, \ldots, i_V$ be the indices for which the coefficients of P change sign, so that we have $\mathrm{sign}\,(a_h) = (-1)^j$ for $h = i_j$ and $\mathrm{sign}\,(a_h) = (-1)^j$ or 0 for any index $h = i_j + 1, \ldots, i_{j+1}$. Write

$$Q(X) = X^{n+1} + b_1 X^n + \cdots + b_k X^{n+1-k} + \cdots + b_n X - c a_0\,.$$

Then, we have the relations $b_k = a_k - c a_{k-1}$ and hence $\mathrm{sign}\,(b_h) = (-1)^j$ for $h = i_j$ ...]

Deduce another proof of the rule of Descartes.

3. Let $f = X^n + a_{n-1}X^{n-1} + \cdots + a_0$ be a monic polynomial with non negative real coefficients bounded above by M. Demonstrate the following implication

$$z \in \mathbb{C} \text{ and } f(z) = 0 \implies \operatorname{Re}(z) \leq 0 \text{ or } |z| \leq \tfrac{1}{2}\big(1 + \sqrt{1 + 4M}\,\big).$$

[When z has a positive real part, demonstrate the following property

$$\Big|1 + \frac{a_{n-1}}{z}\Big| \geq 1 \text{ and } |z| > \tfrac{1}{2}\Big\{1 + \sqrt{1 + 4M}\,\Big\} \implies \sum_{k=0}^{n-2} \big|a_k z^{n-k}\big| < 1.$$

To conclude, consider the quantity $|z^{-n} f(z)|$.]

4. Let f be a polynomial with real coefficients. We consider the real polynomial

$$F(x) = x f'(x) + a f(x),$$

where a is any real number. Let r^+ be the number of strictly positive roots of f, counted with their multiplicities. We want to show that the number of positive real roots of F is at least equal to $r^+ - 1$ (this theorem is due to de Gua).

1°) Let $b_1, \ldots, b_s$ be the positive roots of f, with $b_1 < \ldots < b_s$. The multiplicity of each b_i being equal to r_i. The number of positive roots of f is $r^+ = r_1 + \cdots + r_s$.

Demonstrate that, for $r_i > 1$, b_i is a root of multiplicity $r_i - 1$ of the polynomial F.

2°) Demonstrate that F has a root in each interval $\left]b_i, b_{i+1}\right[$.

[First, demonstrate the following facts :

$$\begin{aligned} x f(x)/f'(x) &\longrightarrow +\infty \quad &\text{if } x \longrightarrow b_i+, \\ x f(x)/f'(x) &\longrightarrow -\infty \quad &\text{if } x \longrightarrow b_{i+1}-, \end{aligned}$$

and f does not change sign on the interval $\left]b_i, b_{i+1}\right[$. Then conclude.]

3°) Deduce that the number of positive real roots of the polynomial F is at least equal to $r^+ - 1$.

4°) Demonstrate that is possible to deduce the rule of Descartes from the theorem of de Gua.

[Here is a way to proceed. Let f be a polynomial with real coefficients and let V be the number of sign variations of the coefficients of f. Finally, let r^+ be the number of positive roots of the polynomial f. We want to demonstrate that $V = r^+ + 2m$, for some nonnegative integer m.

Argue by induction on V. If $V = 0$, the rule of Descartes is obvious : r^+ is zero. If V is positive, and if the first change of sign occurss between the coefficients of the terms x^{n-i} and x^{n-j}, consider the polynomial

$$F(x) = xf(x) - af(x), \quad \text{where} \quad a = n - (i+j)/2.$$

Demonstrate that the number of changes of signs of F is equal to $V - 1$. Deduce the upper bound $r^+ \leq V$. Then, demonstrate that r^+ and V are of the same parity.]

5°) Let r^- be the number of negative roots of f. Demonstrate that the number of negative roots of F is at least equal to $r^- - 1$.

6°) Let r_0 be the order of f in zero. Demonstrate that F has one root of order at least r_0 at the point zero.

7°) Let r be the number of real roots of f. Demonstrate that the number of real roots of F is least equal to $r - 2$. Demonstrate that this number is exactly $r - 2$ in the case of the polynomial $f = x^3 - x$ when we choose $a = -2$.

8°) In this question, we suppose that all the roots of f are real and we call n the degree of f, and suppose $n \geq 1$.

(i) Demonstrate that the roots F are all real when a is strictly positive.

[Suppose that f has at least one root > 0 and let b_1 the smallest of these positive roots. For $\varepsilon > 0$ small enough, demonstrate that the numbers $f(\varepsilon)$ and $F(\varepsilon)$ are of same sign, whereas the numbers $f(b_1 - \varepsilon)$ and $F(b_1 - \varepsilon)$ are of opposite sign. Deduce that F has one root in the open real interval $]0, b_1[$. Conclude ...]

(ii) Demonstrate that the roots of F are all real for $a < -n$.

5. Demonstrate that a polynomial with real coefficients has at least two imaginary roots in any one of the three following cases :

(i) There exist two indices i and j such that $i < j - 1$, for which the coefficients a_i and a_j are nonzero and of the same sign, whereas the a_ks are zero for $i < k < j$.

(ii) There exist two indices i and j such that $i < j - 2$, for which the coefficients a_i and a_j are nonzero and of opposite signs, whereas the coefficients a_k are zero for $i < k < j$.

(iii) The polynomial has three successive coefficients which are in a geometrical sequence.

6. (Euler) If the roots of the polynomial

$$f(x) = a_n x^n + \cdots + a_0$$

are all real, then we have

$$a_i^2 \geq a_{i-1}\, a_{i+1} \quad \text{for} \quad i = 1, 2, \ldots, n.$$

7. This exercise aims to demonstrate the following theorem which is due to M. Riesz :

> Let f be a polynomial with real coefficients, whose roots are all real and simple, and let d be the minimum of the distances between two roots of f. Then, all the roots of the derivative polynomial f' are real and simple, moreover the distance between any two of these roots is always $> d$.

The demonstration we suggest was developed by Stoyanoff ‡.

Consider a polynomial f as in the statement.

1°) Demonstrate that the roots of the derivative polynomial f' are all real and simple and that, if $x_1, \ldots, x_n$ are the roots of f and $y_1, \ldots, y_{n-1}$ those of f', in increasing order, then we have

$$x_1 < y_1 < x_2 < y_2 < \ldots < x_{n-1} < y_{n-1} < x_n .$$

[Refer to Rolle's theorem!]

---

‡ A. Stoyanoff, Sur un théorème de M. Marcel Riesz, *Nouvelles Annales de Mathématique*, 1 (6), 1926, p. 97–99.

2°) Let k be an index such that the difference $y_{k+1} - y_k$ is minimal. We want to demonstrate that the following inequality holds

$$\text{(i)} \qquad y_{k+1} - y_k > d.$$

Show that we may suppose that we have $y_k + d > x_{k+1}$; in others words, that the points $y_k + d$ and y_{k+1} both belong to the interval $\left]x_{k+1}, x_{k+2}\right[$. Show that the function $F(x) = f'(x)/f(x)$ is strictly decreasing in the interval $\left]x_{k+1}, x_{k+2}\right[$ and satisfies $F(y_{k+1}) = 0$.

[Notice the relation $F(X) = \sum_{j=1}^{n} \dfrac{1}{x - x_j} \cdot$]

Deduce that inequality (i) is true if $F(y_k + d)$ is positive and demonstrate that $F(y_k + d)$ is positive.

[Notice the relations

$$\begin{aligned} F(y_k + d) &= F(y_k + d) - F(y_k) \\ &= \sum_{j=1}^{n} \Big(\frac{1}{y_k + d - x_j} - \frac{1}{y_k - x_j} \Big) \\ &= \frac{1}{y_k + d - x_1} + \frac{1}{x_n - y_k} + \sum_{j=1}^{n-1} \frac{x_{j+1} - x_j - d}{(y_k - x_j)\,(y_k + d - x_{j+1})}, \end{aligned}$$

and verify that all the terms of this last sum are ≥ 0, and nonzero.]

8. Let $a_1, \ldots, a_n$ be given real numbers. We consider the polynomial

$$f(X) = (X + a_1) \cdots (X + a_n),$$

and we put

$$f(X) = \sum_{i=0}^{n} p_i \binom{n}{i} X^i,$$

so that

$$p_0 = 1, \quad \text{and} \quad p_i = \frac{\sum a_1 a_2 \cdots a_i}{\binom{n}{i}} \quad \text{for} \quad 1 \leq i \leq n,$$

where the sum is extended to all the products of i terms extracted from the sequence $a_1, \ldots, a_n$.

1°) Let s be an integer, $s \geq 1$. Demonstrate the formula

$$f^{(s)}(X) = n\,(n-1)\ \cdots\ (n-s+1) \\ \times\Big\{X^{n-s} + \cdots + \binom{n-s}{k} p_k X^{n-s-k} + \cdots + p_{n-s}\Big\}.$$

2°) From now on, suppose that the points a_i are pairwise distinct. Demonstrate that all the roots of the polynomial $f^{(s)}$ are real and pairwise distinct. Let k be an integer between 0 and n, deduce that all the roots of the polynomial

$$p_{k+1}Y^{k+1} + \binom{k+1}{1} p_k Y^k + \binom{k+1}{2} p_{k-1} Y^{k-1} + \cdots + 1,$$

are also real and pairwise distinct.

3°) Show that the polynomial

$$p_{k+1}Y^2 + 2p_k Y + p_{k-1}$$

has two distinct real roots.

[Consider the derivative of order $k-1$ of the previous polynomial.]

4°) Deduce the inequality

$$p_k^2 > p_{k-1} \cdot p_{k+1} \,.$$

5°) Suppose, moreover, that the a_i are positive. Demonstrate the sequence of inequalities

$$p_n^{1/n} < \cdots < p_3^{1/3} < p_2^{1/2} < p_1 \,.$$

Deduce, in particular, the well-known inequality between arithmetical mean and geometrical mean :

$$(a_1 \cdots a_n)^{1/n} < \frac{a_1 + \cdots + a_n}{n} \,.$$

9. We consider the number r of distinct real roots of a trinomial with real coefficients

$$f(x) = x^n + px^m + q.$$

1°) Suppose n odd. Demonstrate that , by some suitable change of variable, we may suppose that m is odd too. When this is the case, demonstrate that the number r is equal to one or three and that f has three distinct roots if, and only if, the following condition holds

$$\left(\frac{mp}{n}\right)^n + \left(\frac{mq}{n-m}\right)^{n-m} < 0.$$

What do we get when $n = 3$?

2°) Now, suppose n odd. Demonstrate that, by some change of variable, we may suppose that one of the two following cases occur : (i) m is odd and p is positive, or (ii) m is even.

We call s the sign of the expression

$$\left(\frac{mq}{n-m}\right)^{n-m} - \left(\frac{mp}{n}\right)^n,$$

so that $s = 0$ or ± 1. Demonstrate that, the number r is given by the following formulas

$$r = \begin{cases} 1 - s, & \text{in case (i)}, \\ 2(1-s), & \text{in case (ii)} \end{cases}$$

10. Demonstrate that all the roots of a polynomial of degree n with real coefficients are real if, and only if, the sequence of Sturm associated to this polynomial has n polynomials whose leading coefficients all have the same sign. What is the result for the polynomial $x^3 + px + q$?

11. Consider a sequence $f_0, f_1, \ldots, f_s$ of polynomials with real coefficients which satisfy the following conditions :

(i) we cannot simultaneously have $f_i(c) = f_{i+1}(c) = 0$ with $0 \le i < s$ and c real,

(ii) if c is a real number such that $f_i(c) = 0$, with $0 < i < s$, then, we have the inequality $f_{i-1}(c) \cdot f_{i+1}(c) < 0$,

(iii) f_s has no real root.

Suppose that $f = f_0$ is a polynomial of degree n. Demonstrate that if the relation

$$\mathrm{V}(f_0, f_1, \ldots, f_s\,;\,-\infty) - \mathrm{V}(f_0, f_1, \ldots, f_s\,;\,+\infty) = n$$

holds, then, all the roots of f are real and simple.

12. This exercise aims to demonstrate the rule of Laguerre, which allows us to obtain upper bounds for the real roots of a polynomial with real coefficients. Let

$$f(x) = a_0 x^n + a_1 x^{n-1} + \cdots + a_n$$

be a polynomial with real coefficients of degree equal to n. Put

$$f_0 = a_0,\ f_1 = x f_0 + a_1,\ \ldots,$$
$$f_i = x f_{i-1} + a_i,\ \ldots,\ f_n = x f_{n-1} + a_n = f\,.$$

1°) Demonstrate that, for any real number c, we have the relation

$$f_i(x) = (x - c)\left\{f_0(c)x^{i-1} + \cdots + f_{i-1}(c)\right\} + f_i(c)\,.$$

2°) Suppose that we have $f_i(c) \geq 0$ for $i = 0, 1, \ldots, n$. Demonstrate that the polynomial f satisfies $f(x) > 0$ for $x > c$.

3°) Demonstrate the implication

$$\left\{ f_i(c) \geq 0 \text{ for } 0 \leq i \leq k \right\} \Longrightarrow \left\{ f_i(c') \geq 0 \text{ for } 0 \leq i \leq k \text{ if } c' \geq c \right\}.$$

4°) Deduce that the following procedure allows to find a real number c such that $f_i(c) \geq 0$ for $i = 0, 1, \ldots, n$.

(i) Take c such that $f_1(c) \geq 0$.

(ii) Let k be the smallest integer that satisfies $k = n + 1$ or the two simultaneous conditions $k \leq n$ and $f_k(c) \leq 0$.

(iii) If $k \leq n$ then replace the value c by a number $c' > c$ such that $f_k(c') \geq 0$ and go back to (ii).

(iv) c fits here.

13. This exercise aims to demonstrate the following theorem, due to Jensen :

> Let f be a polynomial with real coefficients and let U be the union of the closed discs whose diameters are the segments joining two imaginary conjugate roots of f, then, any imaginary root of the polynomial f' belongs to the set U.

[Develop the rational fraction $f'(z)/f(z)$ into simple elements and use the following fact : when u is a root of f and z a complex point outside of U, then, we have the relation

$$\operatorname{sign}\left\{\operatorname{Im}\left(\frac{1}{z-u}+\frac{1}{z-\bar{u}}\right)\right\} > 0\cdot]$$

14. Demonstrate the corollary of Proposition 5.7 using only Rolle's theorem.

15. Let f be a polynomial of degree n with real coefficients. For x real, we denote $\mathrm{v}(x)$ the number of changes of signs of the sequence of the values $f^{(i)}(x)$ for i running from 0 to n. Suppose that all the roots of f are real. Demonstrate that when a and b are real, with $a < b$, and a and b are not roots of f, the number of real roots of f lying between a and b is exactly equal to $\mathrm{v}(a) - \mathrm{v}(b)$.

16. Demonstrate the following theorem due to Laguerre : Let

$$f(x) = a_0x^n + \cdots + a_n\,,$$

be a polynomial with real coefficients. For any real number x we put

$$f_0(x) = a_0,\ f_1(x) = xf_0(x) + a_1,\ \ldots$$
$$f_i(x) = xf_{i-1}(x) + a_i,\ \ldots,\ f_n(x) = xf_{n-1}(x) + a_n,$$

and we call $\mathrm{v}(x)$ the number of changes of signs of the sequence of the values of the numbers $f_i(x)$ for i running from 0 to n. Let a and b be two real numbers, $0 < a < b$, which are not roots of the polynomial f. Then, the number r of real roots of f lying between a and b is equal to

$$r = \mathrm{v}(a) - \mathrm{v}(b) - 2m,$$

where m is a nonnegative rational integer.

Generalize this result to functions like

$$F(x) = a_1 x^{\alpha_1} + \cdots + a_n x^{\alpha_n}\,,$$

where the exponents are any real numbers.

17. Determine, in terms of the values of the real parameter a, the number of real roots of the polynomial $X^4 - 3X^2 + 2X + a$.

18. Demonstrate the "rule of Sturm" : Let a be a positive real number such that the sequence of the coefficients of the polynomial $(x-a)\,f(x)$ admits $2k+1$ more changes of sign than that of the coefficients of the real polynomial $f(x)$, then f has at least $2k$ imaginary roots.

[Let p be the number of positive roots of f, let q be the number of its negative roots, and let v be the number of changes of signs of the sequence of the coefficients of the polynomial f; demonstrate the inequalities $p \le v$ and $q + v + 2k + 1 \le n + 1$.]

19. Let $f(X) = \alpha_0 X^n + \cdots + \alpha_n$, be a polynomial with complex coefficients. Put

$$\alpha_j = a_j + i\,b_j, \quad \text{where} \quad a_j \text{ and } b_j \text{ are real,} \quad \text{for } j = 0, 1, \ldots, n,$$

and

$$P(X) = a_0 X^n + \cdots + a_n, \qquad Q(X) = b_0 X^n + \cdots + b_n\,.$$

Suppose that all the zeros of f have a positive imaginary part. Demonstrate that all the zeros of the polynomials P and Q are real.

[Let $z_1, \ldots, z_n$ be the roots of f and let z be a zero of P or of Q. Then, demonstrate the relation

$$|z - \bar{z}_1| \cdots |z - \bar{z}_n| = |z - z_1| \cdots |z - z_n|\,.]$$

20. Let m be a positive integer, and consider n nonzero real numbers $a_1, \ldots, a_n$, $n \geq 2$, and a strictly increasing sequence of n real numbers $x_1, \ldots, x_n$. Demonstrate that the number r of real roots of the polynomial

$$f(x) = a_1\,(x - x_1)^m + \cdots + a_n(x - x_n)^m$$

is equal to $v - 2k$, where k is a nonnegative rational integer and v designates the number of changes of signs of the sequence $a_1, a_2, \ldots, a_n, (-1)^m a_1$.

[Consider the derivative of the rational fraction $f(x)(x - x_n)^{-m}$.]

21. Let P and Q be two polynomials with real coefficients, without common roots and satisfying $\deg(P) - 1 \leq \deg(Q) \leq \deg(P)$.

1°) Demonstrate the equivalence between the two followings properties :

(i) All the roots of the polynomials P and Q are real, and satisfy the following property :

if $x_1, \ldots, x_p$ are the roots of P and if $y_1, \ldots, y_q$ are those of Q, then we have the sequence of inequalities $x_1 < y_1 < x_2 < \cdots$

(ii) For any pair of real numbers a and b not both equal to zero, the polynomial $F(x) = a\,P(x) + b\,Q(x)$ only has real roots.

[To prove the implication (i) $\Longrightarrow$ (ii), notice that $F(x_k) = bQ(x_k)$ and that the $Q(x_k)$ change sign when k increases one unit. Deduce that F has at least $p - 1$ real roots . Conclude.

To prove the implication (ii) $\Longrightarrow$ (i), first, demonstrate that, for z imaginary, the quotient $h(z) = P(z)/Q(z)$ is imaginary too (otherwise F would have an imaginary root).

Deduce that, on one hand, the imaginary part of $h(z)$ keeps a constant sign in any domain where that of z keeps a constant sign too, and, on another hand, that all the roots of the polynomials P and Q are real. Demonstrate that all the residues of the function $h(z)$ must have the same sign. Demonstrate that if y and y' are two consecutive roots of the polynomial Q then $h(y+)$ and $h(y'-)$ have opposite signs, hence $P(x)$ changes sign between the values y and y'.]

2°) Deduce that if f and g are two polynomials having all their roots real and satisfying the property considered in part (i) of the preceding question, then the same property holds for their derivatives.

[Consider the polynomial $F(x) = a\,f(x) + b\,g(x)$ and its derivative.]

Give another demonstration of the theorem of Riesz (see Exercise 7), proceeding as follows : if d is the minimal distance between two distinct roots of f and if $h < d$ then $f(x)$ and $f(x+h)$ have all their roots real and satisfying property $(*)$. Then consider $f'(x)$ and $f'(x+h)$ and use what we have just demonstrated.

22. Let a be a real number and let P be a nonconstant polynomial with real coefficients. Demonstrate that the polynomial $aP(x) + P'(x)$ has no more imaginary roots than the polynomial P itself.

23. Let $x_1, \ldots, x_n$ be real numbers that belong to some real closed interval $[a, b]$. Let r be the number of real zeros $> b$ of the polynomial

$$f(x) = a_0 + a_1(x - \xi_1) + a_2(x - \xi_1)(x - \xi_2) + \\ \cdots + a_n(x - \xi_1)\cdots(x - \xi_n),$$

where the a_is are real numbers. Demonstrate that there exists a nonnegative rational integer m such that $r = v^+ - 2m$, where v^+ designates the number of changes of signs of the sequence $a_0, a_1, \ldots, a_n$. Demonstrate also that there exists a nonnegative rational integer m' such that $r^- = v^- - 2m'$, where v^- designates the number of changes of signs of the sequence a_0, $-a_1, \ldots, (-1)^n a_n$ and r^- the number of roots of the polynomial f that are strictly smaller than α.

24. Demonstrate the following generalization, due to Hurwitz, of the theorem of Budan-Fourier : Let a and b be two real numbers, with $a < b$. Let f be a polynomial which becomes zero neither in a nor in b, and let r be an integer such that $f^{(r)}$, the derivative of order r of f, does not become zero either in a or in b. Let N, and respectively N_r, be the number of zeros (counted with their multiplicities) of f, respectively of $f^{(r)}$, that belong to the interval $[a, b]$. Then, there exists a nonnegative integer m such that

$$N - N_r = \mathrm{v}(a) - \mathrm{v}(b) - 2m,$$

where $\mathrm{v}(x)$ designates the number of changes of signs of the sequence

$$\big(f(x),\, f'(x),\, \ldots,\, f^{(r)}(x)\big).$$

25. Let f be a polynomial with real coefficients, of degree n, and let a and b be two real numbers, $a < b$. Take again the notations of the theorem of Budan-Fourier and suppose that we have

$$N = v(a) - v(b) - 2m.$$

Demonstrate that the polynomial f has at least $2m$ imaginary roots.

[Let N_1 be the number of roots of f strictly smaller than a and N_2 be the number of roots of f which are $> b$. With obvious notations, we have the relation $N + N_1 + N_2 + 2(m + m_1 + m_2) = n \ldots$]

26. Let $P = X^4 + pX^2 + qX + r$ be a polynomial with real coefficients. Find necessary and sufficient conditions on the parameters p, q, and r such that $P(x)$ takes only nonnegative values when x runs over $\mathbb{R}$.

[Hint : First, consider the case $r = 0$. Then, use the change of variable $y = x^{-1}$ and apply Sturm's method.]

27. Let f be a polynomial with real coefficients whose derivative does not become zero on some interval $I = \big[a, b\big]$.

1°) Demonstrate that the function

$$\varphi(x) := x - f(x)/f'(x)$$

is increasing (or respectively decreasing) in any subinterval of I where the quantities f and f'' have the same sign (and respectively are of opposite signs).

2°) We consider the following situation — which we have already met during the proof of Vincent's theorem — where the inequalities

$$f(a) < 0, \quad f^{(i)}(a) \geq 0 \ \text{ for all } i \geq 1 \quad \text{and} \quad f(b) > 0$$

hold. Demonstrate the inequalities

$$b - \frac{f(b)}{f'(b)} < \xi < a - \frac{f(a)}{f'(a)}.$$

3°) We keep the hypothesis of the previous question and we put

$$M = \max\left\{\frac{f''(x)}{2f'(x)}\,;\ a \le x \le b\right\}.$$

Consider the function

$$\psi(x) = \varphi(x) - M(x - x)^2 .$$

Demonstrate that the function ψ is decreasing on the interval I. Deduce the inequalities

$$\xi > b - \frac{f(b)}{f'(b)} - M\,(x - \xi)^2$$

and

$$\xi > b - \frac{f(b)}{f'(b)} - \frac{f''(b)}{2f'(a)}\,(b - a)^2 .$$

28. To a nonconstant polynomial f with real coefficients, we associate the monic polynomial F whose roots are the squares of the differences of the roots of f.

1°) Demonstrate that all the roots of F are real and strictly positive if, and only if, all the roots of the polynomial f are real and distinct.

2°) Deduce that all the roots of f are real and distinct if, and only if, all the coefficients of F are nonzero and of opposite signs for any two consecutive indices. Apply this result to the binomial $x^2 + ax + b$.

29. Demonstrate that the only polynomials A, B, $C \in \mathbb{R}[X]$ such that

$$A^2(X) = X\{B^2(X) + C^2(X)\}$$

are zero polynomials. What happens over $\mathbb{C}[X]$?

30. Let F be a univariate nonconstant polynomial with real coefficients and positive leading coefficient.

1°) Suppose that a is a real root of F such that $F'(x)$ is positive for $x > a$. Let x_0 be a real number, $x_0 > a$. Show that the sequence of numbers defined by Newton's method

$$(*) \qquad x_{n+1} = x_n - \frac{F(x_n)}{F'(x_n)}, \quad n \ge 0,$$

is decreasing and converges to the root a.

2°) Suppose that F has a real positive root r and that the modulus of any complex root of F is bounded above by r. If x_0 is a real number bigger than r, show that the sequence x_n defined by $(*)$ converges to r.

[Hint : Use the theorem of Gauss-Lucas.]

3°) Let R be the polynomial defined by the formula

$$R(Y) = \mathrm{Res}_X\Big(F(X), X^d\, F(Y/X)\Big),$$

where d is the degree of F. Prove that the roots of R are the products $\alpha_i\, \alpha_j$, where the αs are the roots of F.

Put $\rho = \max\{|\alpha_1|, \ldots, |\alpha_d|\}$. Let y_0 be a real number, $y_0 > \rho^2$. Demonstrate that the sequence (y_n) defined by $y_{n+1} = y_n - R(y_n)/R'(y_n)$ converges to ρ^2.

31. Let F be a squarefree polynomial, $F \in \mathbb{R}[X]$. Demonstrate the formula

$$\mathrm{sign}\Big(\mathrm{Discr}\,(F)\Big) = (-1)^s,$$

where $2s$ is the number of nonreal roots of F.

[Let $\alpha_1, \ldots, \alpha_n$ be the roots of the polynomial F, and let Δ be its discriminant. Then,

$$\mathrm{sign}\,(\Delta) = \mathrm{sign}\,\Big(\prod_{1 \le i < j \le n} (\alpha_i - \alpha_j)^2\Big).$$

And notice that, when α is real and β is not then

$$\mathrm{sign}\Big\{(\alpha-\beta)(\alpha-\bar\beta)\Big\} > 0 \quad \text{and} \quad \mathrm{sign}\Big\{(\beta-\bar\beta)^2)\Big\} < 0.]$$

Chapter 6

POLYNOMIALS OVER FINITE FIELDS

The first step of modern algorithms of factorization of polynomials with integer coefficients consists in factorizing their image modulo some prime number. This is the reason why, in this chapter, we study the factorization of polynomials over finite fields. Most of the results of this theory were developed by E.R. Berlekamp.

1. Finite fields

1. General results

In this chapter the letter K designates a finite commutative field*).

The only morphism of $\mathbb{Z}$ in the field K that sends the integer 1 on the unity element of the field K (also written 1) is not injective; its nullspace is equal to $m\mathbb{Z}$, for some positive integer m (proof : this kernel is a subgroup of $\mathbb{Z}$, and apply the corollary of Theorem 1.2).

Hence, we have an injective map $\varphi : \mathbb{Z}/m\mathbb{Z} \longrightarrow K$, which shows that the quotient ring $\mathbb{Z}/m\mathbb{Z}$ is an integral domain. Hence the rational integer m is a prime number, say $m = p$. This prime number p is called the *characteristic* of the field K.

Thus a finite field of characteristic p has a subfield that is isomorphic to the field $\mathbb{Z}/p\mathbb{Z}$ [equal to the image $\varphi(\mathbb{Z}/p\mathbb{Z})$]. The field $\mathbb{Z}/p\mathbb{Z}$ will be written $\mathbb{F}_p$. It is easy to verify that K has a structure of vector space over $\mathbb{F}_p$, consequently — if n is the dimension of this vector space — the field K has p^n elements.

---

* Every finite field is commutative per the theorem of Wedderburn.

If K is a commutative field of characteristic p, if x and y are any two elements of K, the binomial formula of Newton, and the fact that the binomial coefficients $\binom{p}{k}$ are divisible by p for $k = 1, 2, \ldots, p-1$, show that we have

$$(x+y)^p = x^p + y^p .$$

If q is a power of p, say $q = p^n$, then by induction on n we also have

$$(x+y)^q = x^q + y^q .$$

If the field K has q elements, the set K^* of the nonzero elements of K is a (multiplicative) group of cardinality $q-1$. After the theorem of Lagrange (cf. Chapter 1), any element of K^* satisfies $x^{q-1} = 1$, and, since $0^q = 0$, we have $x^q = x$ for each element x of K.

Consequently, K is isomorphic to the splitting field of the polynomial $X^q - X$ over the field $\mathbb{F}_p$. If K and K' are two finite fields of same cardinality q, then they are isomorphic; hence a field with q elements can be written $\mathbb{F}_q$.

Conversely, if q is a power of a prime number p and if Ω_p is a fixed algebraic closure of the field $\mathbb{F}_p$, the set E of the elements of Ω_p that satisfy $x^q = x$ is a subfield of Ω_p : indeed,

$$\begin{aligned}
&x, y \in E \implies (xy)^q = xy && \text{thus} \quad xy \in E,\\
&x \in E \quad \text{and} \quad x \neq 0 \implies x^{-1} \in E,\\
&x, y \in E \implies (x+y)^q = x^q + y^q = x + y && \text{thus} \quad x + y \in E,\\
&x \in E \implies -x \in E .
\end{aligned}$$

The set E has q elements since the roots of the polynomial $X^q - X$ are simple (the derivative of this polynomial is equal to -1). Consequently, E is a field with q elements. In other words, the field $\mathbb{F}_q$ does exist!

By applying Proposition 2.3, we get the following result.

PROPOSITION 6.1. — *Any finite multiplicative subgroup of a commutative field is cyclic. In particular, the multiplicative group $\mathbb{F}_q^*$ is cyclic.*

Let us summarize what has just been demonstrated.

THEOREM 6.1. — *Any finite field K has a cardinality equal to p^n for some positive integer n, where p is a prime number equal to the characteristic of the field K. Conversely, if p is a prime number and n is any positive integer, there exists a finite field, and only one up to an isomorphism, of cardinality $q = p^n$; we denote it by $\mathbb{F}_q$.*

Every element x of the field $\mathbb{F}_q$ satisfies $x^q = x$. Moreover, if K is a field for which $\mathbb{F}_q$ is a subfield, the relation $x^q = x$ characterizes the elements of the field $\mathbb{F}_q$ among those of K. Hence the factorization

$$X^q - X = \prod_{x \in \mathbb{F}_q} (X - x).$$

2. The operations in a finite field

Here we study the arithmetical operations in the finite field $\mathbb{F}_q$. From a mathematical point of view, there is essentially only one finite field with q elements. From the computational point of view, the situation is different : to compute with the elements of $\mathbb{F}_q$ we must choose a representation of this field and carrying out the computations in $\mathbb{F}_q$ depends very much on this representation. A very detailed study of this important question can be found in the exellent book by E.R. Berlekamp†.

If $q = p$, the representation of $\mathbb{F}_q$ creates no problem. Suppose that $q = p^n$ with $n \geq 2$. To simplify, we only consider the two following representations :

(i) $\mathbb{F}_q$ is a $\mathbb{F}_p$–vector space of dimension n. We choose some basis $\{b_1, \ldots, b_n\}$ of this space and we represent each element of $\mathbb{F}_q$ by its components on this basis.

(ii) We choose one element x of $\mathbb{F}_q^*$ which generates this group. Then any nonzero element $\mathbb{F}_q$ is written in one, and only one, way as x^m, with $0 \leq m < q - 1$.

In representation (i), the addition is done very simply, component by component; it is equivalent to n additions modulo p. However, representation (ii) is not well adapted to addition. There is no simple rule allowing

† *Algebraic Coding Theory*, McGraw-Hill, New-York, 1968.

us to determine an integer m such that $x^m = x^h + x^k$, when h and k are given (such that the sum $x^h + x^k$ is nonzero). It seems *a priori* that we have to build a table of $q(q-1)/2$ elements. Nevertheless, if, for example, we suppose the inequality $k \geq h$, we can write

$$x^h + x^k = x^h(1 + x^{k-h})$$

and only compute once the table of the q elements $1+x^j$, for the successive exponents $j = 0, 1, \ldots, q-2$.

The computation of the opposite of an element is made very simply in the two representations. It is clear for the first representation. For the second representation, we notice that there is no problem in characteristic two, since, in this case, we just have $-y = y$. Whereas, for q odd, we use the fundamental formula

$$x^{(q-1)/2} = -1,$$

which is the result of the two relations $\{x^{(q-1)/2)}\}^2 = 1$ and $x^{(q-1)/2} \neq 1$. Indeed, when $y = x^k$, this formula allows us to write

$$-y = x^j, \quad \text{where} \quad j = \big(k + (q-1)/2\big) \bmod (q-1).$$

Representation (ii) is very convenient to compute the product : we just have

$$x^h \cdot x^k = x^j, \quad \text{where} \quad j = (h+k) \bmod (q-1).$$

However, to compute a product in representation (i) we need to know the values of the different products $b_i \cdot b_j$. It seems better to choose a basis like $\{1, z, \ldots, z^{n-1}\}$ — in other words, to choose the isomorphism

$$\mathbb{F}_q \sim \mathbb{F}_p[X]/(P(X)),$$

where P is the minimal polynomial of some element z over the field $\mathbb{F}_p$ (notice that P is of degree n when $\{1, z, \ldots, z^{n-1}\}$ is a basis). In this case, we only have to compute once the representations on this basis of the $n-1$ powers $z^n, \ldots, z^{2n-2}$ (we shall prove later on that such an element z always exists). Then, we have only to compute n^2 products in $\mathbb{F}_p$ and at most n^2 additions.

The computation of the inverse is trivial in representation (ii) : if y is equal to x^k, its inverse y^{-1} is equal to x^{q-1-k}.

To compute y^{-1} in the case of representation (i), we can use the relation $y^{-1} = y^{q-2}$. The right-hand side can be computed by only $\mathrm{O}(\operatorname{Log} q)$ products. As we have already seen, this method can also be used for the prime field $\mathbb{F}_p$. In the case of the representation $\mathbb{F}_q \sim \mathbb{F}_p[X]/(P(X))$, we can compute the inverse with the algorithm of Euclid : indeed, if a nonzero element x of $\mathbb{F}_q$ is represented by the polynomial $A(z)$ and if we have computed a relation of Bézout

$$AU + PV = 1,$$

then the inverse of x is simply equal to $U(z)$.

3. Determination of a primitive element of $\mathbb{F}_q^*$

A primitive element x of $\mathbb{F}_q^*$ is characterized by the fact that it does not satisfy any of the relations $x^r = 1$, where r runs over the strict divisors of $q-1$. Thus, a nonzero element x of the field $\mathbb{F}_q$ is a primitive element of this field if, and only if, it satisfies the condition

$$s \mid (q-1) \text{ and } s \text{ prime} \implies x^{(q-1)/s} \neq 1.$$

Hence the primitive elements of the multiplicative group $\mathbb{F}_q^*$ are exactly the roots of the polynomial

$$\Phi_{q-1}(X) = (X^{q-1} - 1)/S(X),$$

where we have put

$$S(X) = \text{g.c.d.}\left\{X^{q-1} - 1, \prod_{r \mid q-1\,;\, 1 \le r < q-1} (X^r - 1)\right\}.$$

The polynomial $\Phi_{q-1}(X)$ is called *cyclotomic polynomial* of order $q-1$.

Since the group $\mathbb{F}_q^*$ is cyclic and of cardinality $q-1$, it contains exactly $\varphi(q-1)$ generators. Hence, the proportion of primitive elements in the group $\mathbb{F}_q^*$ is the quotient $\varphi(q-1)/(q-1)$. It is known that Euler's function satisfies $\varphi(n)/n \ge c\,(\operatorname{Log}\operatorname{Log} n)^{-1}$, for some positive constant c (see Exercise 1). Hence the fact that, by successively taking k elements at random in the set $\mathbb{F}_q^*$, until we find a primitive element, we "almost always" have to make only $k = \mathrm{O}(\operatorname{Log}\operatorname{Log} n)$ tries.

4. Determination of z such that $\mathbb{F}_q = \mathbb{F}_p[z]$

We, already, know that every nonzero element of $\mathbb{F}_q$ satisfies $z^{q-1} = 1$. Thus, an element z such that $\mathbb{F}_q = \mathbb{F}_p[z]$ satisfies $z^{q-1} = 1$, but it must not satisfy any relation $z^{q'-1} = 1$, where $q' = p^d$ and where d is a strict divisor of n (otherwise, z belongs to the subfield $\mathbb{F}_{q'}$ of $\mathbb{F}_q$).

Hence the sought elements z are exactly the roots of the polynomial

$$P(X) = (X^{q-1} - 1)/R(X),$$

where

$$R(X) = \text{g.c.d.}\left\{X^{q-1} - 1, \prod_{d|n\,;\,1 \le d < n} (X^{p^d - 1} - 1)\right\}.$$

Moreover, each of these elements z is the root of a polynomial of $\mathbb{F}_p[X]$ irreducible and of degree n. Thus, $P(X)$ is equal to the product of the polynomials of $\mathbb{F}_p[X]$, which are irreducible and of degree n. If Q designates any one of these polynomials, the field $\mathbb{F}_q$ is isomorphic to the quotient ring

$$\mathbb{F}_p[X]/(Q(X)).$$

To determine one of these polynomials Q, we, obviously only have to factorize the polynomial P in the field $\mathbb{F}_p$; but it is easier — when q is small enough — to proceed as follows :

- First, compute all the possible products of a polynomial of degree n' over $\mathbb{F}_p$ by a polynomial of degree $n - n'$ over $\mathbb{F}_p$.

- The previous computation gives the list of the reducible polynomials of degree n, hence a polynomial Q is any polynomial of degree n that does not appear in this list.

It is also possible to choose an arbitrary polynomial of degree n and to test if it divides P; if it is the case, the polynomial is irreducible, otherwise we try another polynomial of degree n ...

As we shall soon see, the number I_n of polynomials of $\mathbb{F}_p[X]$ irreducible and of degree n is equivalent to p^n/n. Hence, each of the previous tries has a "chance" to succeed close to $1/n$; for k consecutive tries, the "chance" to fail is close to $(1 - 1/n)^k$. Hence, "in general" the maximum number of tries is $\mathrm{O}(n)$.

5. An example : $\mathbb{F}_8$

We are going to determine $\mathbb{F}_8$, "the" field with eight elements. If z is any element of $\mathbb{F}_8$ other than 0 or 1 (notice that 0 and 1 are the two elements of the prime field $\mathbb{F}_2$), then $\mathbb{F}_8 = \mathbb{F}_2[z]$.

The first step consists of determining the possible minimal polynomials for z, that is the irreducible polynomials of degree three over $\mathbb{F}_2$. It is simpler to notice that the polynomials of degree one or two for which zero is not a root are

$$X+1, \quad X^2+1, \quad X^2+X+1.$$

Hence, the reducible polynomials of degree three with nonzero constant term are

$$X^3+X^2+X+1, \quad X^3+1;$$

which shows that the polynomials of degree three irreducible over the prime field $\mathbb{F}_2$ are

$$P(X) = X^3+X+1, \quad Q(X) = X^3+X^2+1.$$

Morever, we verify that

$$X^8 - X = X(X-1)P(X)Q(X).$$

Let us take, for example, for z a root of the polynomial $P(X)$. The application $x \mapsto x^2$ keeps the polynomial P invariant, but permutes its roots. The two other roots of P are, hence z^2 and $z^4 = z^2 + z$ (because of the relation $z^3 + z + 1 = 0$). The table of multiplication is given by the identities

$$z^4 = z^2 + z, \; z^5 = z^3 + z^2 = z^2 + z + 1, \; z^6 = z^3 + z^2 + z = z^2 + 1.$$

We have $z^7 = 1$ and z generates $\mathbb{F}_8^*$ (as any element of $\mathbb{F}_8^*$ other than 1, since this group is of prime cardinality). The cyclotomic polynomial Φ_7 is equal to

$$\Phi_7(X) = X^6 + X^5 + X^4 + X^3 + X^2 + X + 1,$$

and this polynomial factorizes over the field $\mathbb{F}_8^*$ as follows

$$\Phi_7(X) = P(X)\,Q(X).$$

2. Statistics on $\mathbb{F}_q[X]$

1. The function of Möbius

The function of Möbius is defined by the formula

$$\mu(n) = \begin{cases} (-1)^k, & \text{if } n = p_1 \cdots p_k,\ k \geq 0,\ p_i \text{ distinct primes,} \\ 0, & \text{otherwise.} \end{cases}$$

This function satisfies the property

$$(*) \qquad \sum_{d|n} \mu(d) = \begin{cases} 1, & \text{if } n = 1, \\ 0, & \text{otherwise.} \end{cases}$$

Indeed, let us call $S(n)$ the left-hand side, if $n = n' n''$, where n' and n'' are relatively prime. Then, we have the relation of "multiplicativity"

$$S(n' n'') = \sum_{d|n''} \mu(d) = \sum_{d'|n'\ ;\ d''|n''} \mu(d') \cdot \mu(d'') = S(n') \cdot S(n''),$$

which proves that it is only necessary to verify formula $(*)$ when n is a power of a prime number; in this case, we have

$$S(1) = 1 \quad \text{and} \quad S(p^k) = 1 - 1 = 0 \quad \text{for} \quad k \geq 1.$$

Now deduce the *Möbius inversion formula.*

PROPOSITION 6.2. — *Consider two functions f and g defined over the set $\mathbb{N}^*$ of positive rational integers, with values in a commutative group G (the law of which is denoted additively). Suppose that f and g are bound by the relation*

$$f(n) = \sum_{d|n} g(d);$$

then

$$g(n) = \sum_{d|n} \mu(d) f(n/d).$$

Proof

Indeed, we have

$$\sum_{d|n} \mu(d) f(n/d) = \sum_{d|n} \mu(d) \sum_{d' \mid (n/d)} g(d') = \sum_{d,d' \,;\, dd' | n} \mu(d) g(d')$$
$$= \sum_{d'} g(d') \sum_{d \mid (n/d')} \mu(d) = g(n)\,.$$

Hence, we have the result. ∎

Corollary. — *Let $\Phi_n(X) \in \mathbb{Z}[X]$ be the cyclotomic polynomial of order n, that is the minimal polynomial of a primitive n-th root of unity. Then $\Phi_n(X)$ satisfies by the formula*

$$\Phi_n(X) = \prod_{d|n} (X^{n/d} - 1)^{\mu(d)}.$$

Proof

We take for G the multiplicative group $\mathbb{Q}(X)^*$ (of nonzero rational functions over the field of rational integers) and use the relation

$$X^n - 1 = \prod_{d|n} (X^{n/d} - 1)\,.$$

Now, the result is obvious. ∎

2. Counting irreducible polynomials

Remember that I_n is the number of unitary irreducible polynomials of degree n with coefficients in the field $\mathbb{F}_q$.

By classifying the elements of $\mathbb{F}_{q'}$, where $q' = q^n$, in function of their degree over $\mathbb{F}_p$, we get

$$q^n = \sum_{d|n} d\, I_d\,.$$

An application of the inversion formula of Möbius leads to the following result.

Proposition 6.3. — *The number I_n of irreducible monic polynomials of degree n with coefficients in the finite field $\mathbb{F}_q$ is given by the formula*

$$I_n = \frac{1}{n}\sum_{d|n} \mu(d)\, q^{n/d}.$$

Hence the inequalities

$$q^n - 2q^{n/2} < nI_n \leq q^n,$$

where the second inequality is strict for $n \geq 2$.

3. Number of squarefree polynomials

The number of squarefree polynomials over a finite field is given by the following proposition.

Proposition 6.4. — *The number of squarefree monic polynomials over the finite field $\mathbb{F}_q$, of degree $d \geq 2$, is equal to $q^d - q^{d-1}$.*

Proof

We only give the principle of the demonstration. It is easy to see that the characteristic function of the monic polynomials over the field $\mathbb{F}_q$ is equal to $(1 - qT)^{-1}$. Hence, the characteristic function of their squares is the rational function $(1 - qT^2)^{-1}$.

Since there is only one way to write any monic polynomial as the product of a squarefree monic polynomial and of the square of a monic polynomial, the characteristic function of the monic and squarefree polynomials is equal to

$$\frac{1}{1-qT}\cdot\left(\frac{1}{1-qT^2}\right)^{-1} = \frac{1-qT^2}{1-qT}.$$

Hence, we have the result. $\square$

4. Study of the number of irreducible factors of a polynomial

The results of this paragraph are taken from an article by Mignotte and Nicolas*. But, here, the constants are explicitly computed.

We call E_n the set of the monic polynomials of degree n belonging to the ring $\mathbb{F}_q[X]$. Hence, we have $\operatorname{Card}(E_n) = q^n$. If a polynomial F belonging to the set E_n admits the decomposition in product of irreducible factors

$$F = {P_1}^{a_1} \cdots {P_k}^{a_k},$$

we set $\omega(F) = k$, in other words

$$\omega(F) = \sum_{P|F\,;\,P \text{ irreducible}} 1.$$

LEMMA 6.1. — *Let T be an integer such that $1 \le T \le n$. For F belonging to E_n, we put*

$$\omega_T(F) = \sum_{P|F\,;\,P \text{ irr.}\,;\,\deg P \le T} 1,$$

thus $\omega_T(F)$ is the number of irreducible distinct factors of F whose degree is at most equal to T [in particular, if $F \in E_n$, then $\omega_n(F) = \omega(F)$]. Then,

$$\sum_{F \in E_n} \omega_T(F) = q^n\,(\operatorname{Log} T - c), \qquad \textit{with } -1 < c < 2.5.$$

Proof

We have

$$\begin{aligned}\sum_{F \in E_n} \omega_T(F) &= \sum_{F \in E_n} \; \sum_{P|F\,;\,P \text{ irr.}\,;\,\deg P \le T} 1 \\ &= \sum_{i=1}^{T} \sum_{\deg P = i} \; \sum_{F\,;\,P|F} 1 \\ &= \sum_{i=1}^{T} I_i q^{n-i},\end{aligned}$$

where P designates an irreducible monic polynomial and where F runs over the set E_n.

---

* M. Mignotte et J.L. Nicolas, Statistiques sur $\mathbb{F}_q[X]$, *Ann. de l'Inst. Henri Poincaré*, 19 (2), 1983, p. 113–121.

From Proposition 6.3, we get

$$I_i = \frac{q^i}{i} - R_i\, q^{i/2}, \qquad \text{with } 0 \le R_i \le \frac{2}{i}\cdot$$

Hence

$$\sum_{F\in E_n} \omega_T(F) = q^n \left(\sum_{i=1}^{T} \frac{1}{i} - \sum_{i=1}^{T} R_i q^{-i/2} \right) = q^n(\operatorname{Log} T - c),$$

where

$$c = \operatorname{Log} T - \sum_{i=1}^{T} \frac{1}{i} + \sum_{i=1}^{T} R_i q^{-i/2}.$$

It is easy to estimate c. On one hand, we have

$$\operatorname{Log} T < \sum_{i=1}^{T} \frac{1}{i} < 1 + \operatorname{Log} T \quad \text{(consider the integral } \textstyle\int_1^T \operatorname{Log} t\, dt\text{)}$$

and, on the other hand,

$$0 < \sum_{i=1}^{T} R_i q^{-i/2} < 2 \sum_{i=1}^{T} \frac{q^{-i/2}}{i} < 2 \sum_{i=1}^{T} \frac{2^{-i/2}}{i} = -2 \operatorname{Log}\left(1 - \tfrac{1}{\sqrt{2}}\right) < 2.456.$$

Hence we have reached the conclusion. □

PROPOSITION 6.5. — *With the notations of the previous lemma, we have*

$$\sum_{F\in E_n} \left\{\omega_T(F) - \operatorname{Log} T\right\}^2 < 2q^n(1 + \operatorname{Log} T),$$

for any integer T, where $2 \le T \le n$.

Proof

We follow the demonstration of Hardy and Wright[‖] for the function ω relative to integers.

‖ *An Introduction to the Theory of Numbers*, Oxford, at the Clarendon Press, 1960.

The quantity $\omega_T(F)\,(\omega_T(F)\,-\,1)$ is the number of pairs (P,Q) of distinct irreducible polynomials such that P and Q divide F [considering that $(P,Q) \neq (Q,P)$]. Hence, we have

$$\omega_T(F)\left\{\omega_T(F)-1\right\} = \sum_{P\neq Q\ ;\ \deg\, P,Q\leq T\ ;\ PQ|F} 1,$$

which implies that

$$\begin{aligned}\sum_{F\in E_n} \omega_T(F)\ (\omega_T(F)-1)\ &\leq \sum_{\deg\, P\leq T\ ;\ \deg P\cdot Q\leq n}\ \sum_{F\ ;\ PQ|F} 1\\ &\leq \sum_{i=1}^{T} I_i\ \sum_{j=1}^{T} I_j\, q^{n-i-j}\\ &\leq\ q^n\ \Big(\sum_{i=1}^{T}\frac{1}{i}\Big)^2 <\ q^n\ (1+\mathrm{Log}\ T)^2.\end{aligned}$$

The demonstration of the proposition ends by using the formula

$$\left\{\omega_T(F)-\mathrm{Log}\ T\right\}^2 = \omega_T(F)\left\{\omega_T(F)-1\right\}+(1-2\,\mathrm{Log}\ T)\,\omega_T(F)+(\mathrm{Log}\ T)^2,$$

as well as Lemma 6.1. ☐

Corollary (Inequality of Tchebytcheff for the function ω). — *For any integer $n\geq 3$ and any positive real number λ, we have the inequality*

$$\mathrm{Card}\left\{F\in E_n\ ;\ \left|\omega(F)-\mathrm{Log}\ n\right|\geq \lambda\sqrt{\mathrm{Log}\ n}\right\}\leq 3\,q^n\lambda^{-2}.$$

Proof

Apply Proposition 6.4 for $T = n$. ☐

3. Factorization into a product of squarefree polynomials

In this section, K is any commutative field, finite or not.

1. Definitions and generalities

Let K be any commutative field and let F be a nonzero polynomial with coefficients in K. We say that F is *squarefree* if F has only simple roots in any field that contains K.¶ Let F' be the derivative of the polynomial F and let Q be the monic g.c.d. of the polynomials F and F'. If this g.c.d. is equal to 1, then Bézout relation for F and F',

$$UF + VF' = 1,$$

shows that F and F' have no common root and thus that F has only simple roots. In other words F is squarefree. Conversely, if F is not squarefree and possesses, in some field L, which contains K, the decomposition

$$F = \lambda P_1^{e_1} \cdots P_k^{e_k},$$

where the P_is are monic polynomials over L, two-by-two distinct, with $e_1 \geq 2$, then, the derivative F' is equal to

$$F' = \lambda {P_1}^{e_1 - 1} \cdots {P_k}^{e_k - 1} R,$$

and the polynomial

$${P_1}^{e_1 - 1} \cdots {P_k}^{e_k - 1}$$

divides Q. Thus, P_1 divides Q, and F and F' are not coprime.

The polynomial R is given by the formula

$$R = \sum_{i=1}^{k} e_i \, P_1 \cdots P_{i-1} \, P_{i+1} \cdots P_k \,,$$

and the quotient F/Q divides the product $P_1 \cdots P_k$; thus this quotient is a squarefree polynomial. We have just proved the following result.

Proposition 6.6. — *Let F be a nonconstant polynomial, with coefficients in some field K. Then F is squarefree if, and only if, F is relatively prime with its derivative. Moreover, the polynomial $G = F/\,\text{g.c.d.}\,(F, F')$ is always squarefree. Besides, if the derivative F' is nonzero and if moreover $\text{g.c.d.}\,(F, F') \neq 1$, then G is a nontrivial divisor of F.*

¶ Equivalent definition : F is squarefree if it is not divisible by the square of a nonconstant polynomial over K. Exercise : prove this equivalence.

2. Case of characteristic zero

When K is of characteristic zero, the preceding formula for R shows that none of the polynomials P_is divides the polynomial R. Thus, in the present case, we have

$$Q = {P_1}^{e_1-1} \cdots {P_k}^{e_k-1} \quad \text{and} \quad F/Q = P_1 \cdots P_k .$$

3. Case of nonzero characteristic

Now, suppose that K is of characteristic p. Then, the formula for the polynomial R shows that some polynomial P_i divides R if, and only if, the integer p divides the exponent e_i.

- If p does not divide all the exponents e_i, then, for example, it divides the first exponents e_1, …, e_h but does not divide any of the exponents e_{h+1}, …, e_k, for some index h, with $0 \leq h < k$. In that case, the derivative F' is different from zero and the polynomial Q is given by the formula

$$Q = {P_1}^{e_1} \cdots {P_h}^{e_h} {P_{h+1}}^{e_{h+1}-1} \cdots {P_k}^{e_k-1}.$$

Then, we have

$$F/Q = P_{h+1} \cdots P_k .$$

- The case where p divides all the exponents e_i occurs if, and only if, the derivative F' of the polynomial F is zero. If $F = \sum a_i X^i$ and if F has a zero derivative, then p divides each index i for which a_i is nonzero. Let p^e be the greatest power of the number p such that p^e divides the g.c.d. of the integers i for which a_i is nonzero. Then $e \geq 1$ and the polynomial F can be written as

$$F = H(X^{p^e}), \qquad \text{with} \quad H \in K[X].$$

Conversely, if the polynomial F is given by such a formula, then its derivative is equal to zero.

Consider the particular case where the field K is equal to the field $\mathbb{F}_q$ and where F is given by the preceding formula. Let u and v be two integers such that the relation of Bézout

$$up^e + v(q-1) = 1, \qquad \text{with} \quad 1 \leq u < q,$$

holds. Then, for every element x of $\mathbb{F}_q$, we have

$$(x^u)^{p^e} = x^{1-v(q-1)} = x;$$

which means that x^u is a root of order p^e of x.

In such a way, we easily obtain a polynomial H_1 such that

$$H(X^{p^e}) = H_1(X)^{p^e}.$$

Indeed, this reduction needs at most $\mathrm{O}(dp^e \operatorname{Log} q)$ multiplications in the field $\mathbb{F}_q$, where d is the degree of H.

Using Proposition 6.6 and the above study, we get the following result.

PROPOSITION 6.7. — *Let K be a finite field or a field of characteristic zero, then every irreducible polynomial of $K[X]$ is squarefree. Thus, the derivative of such a polynomial is nonzero.*

Remark

If $K = \mathbb{F}_p[Y]$ and $F(X) = YX^p - 1$, then $F' = 0$, nevertheless F is irreducible over K (prove this remark as an exercise).

4. Algorithms of decomposition into a product of squarefree polynomials

In this section, K is a field of characteristic zero or a finite field, the characteristic of which is denoted p. Let F be a nonconstant polynomial of $K[X]$. To obtain a decomposition of F into a product of squarefree polynomials, we can proceed as follows :

(i) Compute the derivative F' of F.

(ii) If $F' \neq 0$, compute the g.c.d. Q of F and of F'. Then, if $Q = 1$, the polynomial F is squarefree. Otherwise, $F = QR$, where R is squarefree and satisfies $0 < \deg R < \deg F$. Then, apply the procedure to the polynomial Q.

(iii) If $F' = 0$, it is possible to compute a polynomial $H \in \mathbb{F}_q[X]$, such that

$$F = H^{p^e}, \quad e \geq 1,\ H' \neq 0;$$

then apply the procedure to the polynomial H.

If d designates the degree of the polynomial F, note that step (i) needs at most d multiplications, whereas the computation of Q can be done in at most $\mathrm{O}(d^2)$ operations over the field K. Finally, the third step can only occur when the characteristic of K is nonzero; moreover, this step needs at most $\mathrm{O}(d/p \cdot \mathrm{Log}\, q)$ multiplications, where q is the cardinality of the field K.

The article of Yun* contains the three following methods to decompose a polynomial into a product of squarefree polynomials.

Let $F \in K[X]$ be a given polynomial. The algorithm computes squarefree polynomials, two-by-two relatively prime, P_1, P_2, ..., P_k, such that F is decomposed into the product

$$F = P_1 {P_2}^2 \cdots {P_k}^k ,$$

with $k \geq 1$ and $P_k \neq 1$.

The first method given was proposed by Tobey and Horowitz.

```
C_1 := g.c.d.(F, F') ; D_1 := P/C_1 ; i := 1 ;
while C_i ≠ 1 do
  begin
  |   C_{i+1} := g.c.d.(C_i, C_i') ; D_{i+1} := C_i/C_{i+1} ;
  |   P_i := D_i/D_{i+1} ; i := i + 1
  end.
```

method 1 (Tobey and Horowitz)

In characteristic zero, the computation is as follows :

$$C_1 = P_2 {P_3}^2 \cdots {P_k}^{k-1} ;\ D_1 = P_1 P_2 \cdots P_k ;\ \ldots$$

and

$$C_i = P_{i+1} {P_{i+2}}^2 \cdots {P_k}^{k-i} ;\ D_i = P_i P_{i+1} \cdots P_k ;\ \ldots$$

* D.Y.Y. Yun, On squarefree decomposition Algorithms, in *Proc. of the* 1976 *AMS Symp. on Symb. and Alg. Comp.*, Yorktown, ed. by R.D. Jenks.

The second method is due to D. Musser.

```
C_1 := g.c.d.(F, F') ; D_1 := P/C_1 ; i := 1 ;
while D_i ≠ 1 do
  begin
  |   D_{i+1} := g.c.d.(C_i, D_i) ; C_{i+1} := C_i/D_{i+1} ;
  |   P_i := D_i/D_{i+1} ; i := i + 1
  end.
```

method 2 (D. Musser)

In characteristic zero, the computation is as follows :

$$C_1 = P_2 {P_3}^2 \cdots {P_k}^{k-1} \; ; \; D_1 = P_1 P_2 \cdots P_k \; ; \; \ldots \; ;$$

$$C_i = P_{i+1} {P_{i+2}}^2 \cdots {P_k}^{k-i} ; \; D_i = P_i P_{i+1} \cdots P_k \; ; \; \ldots$$

The third method was given by Yun.

```
Q := g.c.d.(F, F') ; C_1 := P/G ; D_1 := F' − C'_1 ; i := 1 ;
while C_i ≠ 1 do
  begin
  |   P_i := g.c.d.(C_i, D_i) ; C_{i+1} := C_i/P_i ;
  |   D_{i+1} := D_i/P_i − C_{i+1}' ; i := i + 1
  end.
```

method 3 (D. Yun)

- In characteristic zero, the computation is now the following :

$$Q = P_2 {P_3}^2 \cdots {P_k}^{k-1} ; \; C_1 = P_1 P_2 \cdots P_k \; ;$$

$$D_1 = \sum_{i=1}^{k} (i-1) P_i{}' \prod_{j \neq i} P_j = P_1 \cdot \Big(\sum_{i=2}^{k} (i-1) P_i{}' \prod_{1 < j \neq i} P_j \Big) ;$$

$$C_2 = P_2 P_3 \cdots P_k \; ; \; \ldots \; ;$$

$$C_h = P_h P_{h+1} \cdots P_k \; ;$$

$$D_h = \sum_{i=h}^{k} (i-h) P_i{}' \prod_{h < j \neq i} P_j = P_h \cdot \Big(\sum_{i=h+1}^{k} (i-1) P_i{}' \prod_{h < j \neq i} P_j \Big) ; \ldots$$

• But, none of these three algorithms works correctly in nonzero characteristic. Consider, for example, the polynomial $F(X) = X^3 + X^2$ over the field $\mathbb{F}_2$. The first procedure computes

$$F' = X^2\,;\; C_1 = X^2\,;\; D_1 = X+1\,;$$

then, it gives indefinitely $C_i = X^2 \ldots$

The second algorithm runs as follows :

$$C_1 = X^2\,;\; D_1 = X+1\,;\; D_2 = 1\,;\; C_2 = X^2$$

and gives $F = (X+1)(X^2)$, which is not the wanted decomposition.

The third algorithm computes successively

$$Q = X^2\,;\; C_1 = X+1\,;\; D_1 = 0\,;\; C_2 = 1$$

and, as the preceding one, gives $F = (X+1)(X^2)$.

The algorithm given at the beginning of this section runs as follows :

$$Q = X^2\,;\; R = X+1\,;\; Q' = 0\,;\; Q = (X)^2\,,$$

and gives $F = (X+1)(X)^2$, which is the correct solution.

The study of costs, realized by D. Yun (*loc. cit.*) for the methods 1 to 3, proves that, for a field of zero characteristic, the second method is more economical than the first one, but less than the third one.

4. Factorization of polynomials over a finite field

Let F be a monic polynomial, of degree d and with coefficients in the field $K = \mathbb{F}_q$, that we want to factorize. Thanks to the procedure of the previous paragraph, we only have to considerate the case where F is squarefree, say $F = P_1 \cdots P_k$, where the P_i are irreducible monic polynomials two-by-two distinct, with $\deg(P) = d_i \geq 1$.

The following method of factorization is mainly due to Berlekamp.

1. Determination of the number of irreducible factors

The main point of this method is the study of the structure of the quotient algebra

$$A = K[X]/\bigl(F(X)\bigr).$$

After the Chinese remainder theorem, this algebra is isomorphic to the product

$$\prod_{i=1}^{k} K[X]/(P_i(X)),$$

but, since each P_i is an irreducible polynomial of degree d_i, we also have the isomorphisms

$$K[X]/(P_i(X)) \sim K_i, \quad \text{for } 1 \le i \le k,$$

where the field K_i is a field of degree d_i above K and hence

$$K_i \sim \mathbb{F}_{q_i}, \quad \text{where} \quad q_i = q^{d_i}.$$

In summary,

$$A \sim K_1 \times \cdots \times K_k.$$

Let $L : A \longrightarrow A$ be the function that associates the element $\alpha^q - \alpha$ to an element α of the algebra A. Since the field K, and (therefore) also the algebra A, is of characteristic p, the application L is linear (the demonstration was given in Section 1.1).

By the isomorphism

$$A \sim K_1 \times \cdots \times K_k,$$

the function L induces on each field K_i the function

$$L_i : x \longmapsto x^q - x.$$

After Theorem 6.1, the nullspace of L_i is equal to $\mathbb{F}_q$. Thus, the nullspace of L satisfies the relation

$$\text{Ker } L \sim K^k,$$

and, in particular, the dimension of this nullspace gives the number of irreducible factors of F, that is

$$\dim_K(\text{Ker } L) = k.$$

To determine the number of irreducible factors of the polynomial F over the field $\mathbb{F}_q$ we only have to compute the dimension of the nullspace of the function L. Then, we have the equivalence

$$F \text{ irreducible} \iff \dim_K(\text{Ker } L) = 1,$$

which gives a quick way to test the irreducibility of a polynomial of $\mathbb{F}_q[X]$.

More generally, let us consider, for $h \geq 1$, the function from A into itself

$$L_h : \alpha \longmapsto \alpha^{q^h} - \alpha.$$

Again, it is a linear function, and it induces one each field K_i the function

$$L_{h,\,i} : x \longmapsto x^{q^h} - x.$$

The nullspace of $L_{h,\,i}$ is isomorphic to the intersection $K_i \cap K^{(h)}$, where $K^{(h)}$ designates the field with q^h elements. Hence, the kernel of the linear map $L_{h,\,i}$ is isomorphic to the field with $q^{(d_i,h)}$ elements [as in Chapter 1, the notation (d_i, h) designates the g.c.d. of d_i and h].

In conclusion, we have the formula

$$\dim_K(\text{Ker } L_h) = \sum_{i=1}^{k} (d_i, h).$$

For example, for $h = 2$, we get the relation

$$\begin{aligned}\dim_K(\text{Ker } L_2) = &\#(\text{irr. factors of } F)\\ &+ \#(\text{irr. factors of } F \text{ of even degree}),\end{aligned}$$

where the symbol # is an abbreviation for "number of".

The knowledge of the dimension of the nullspace of the linear function L_h provides precious information on the type of factorization of the polynomial F.

For the computation, we represent the function L_h by the matrix $Q_h - I$, where Q_h is the matrix $d \times d$ over K whose ith line represents the element

$$X^{q^h\,(i-1)} \bmod F(X)$$

on the basis $(1, X, \ldots, X^{d-1})$ of the algebra A. The computation of this matrix Q_h is done by a number of operations in K at most $\mathrm{O}\left(h\, d^2 \operatorname{Log} q\right)$.

If we take a square matrix M of size $n \times n$ over a field, the usual algorithm of triangularization of M leads to the determination of the rank of this matrix and gives a basis of the nullspace which associated to it in $\mathrm{O}(n^3)$ operations†. Hence we can compute the dimension of the nullspace of L_h and a basis of this space in at most $\mathrm{O}(d^3)$ operations. In particular, we can test if F is irreducible over K in a total number of $\mathrm{O}(d^3 + d^2 \operatorname{Log} q)$ operations in the field K.

2. Complete factorization

The key to the algorithm of Berlekamp is the following result, whose demonstration is based on the decomposition of the algebra A,

$$A \sim K_1 \times \cdots \times K_k,$$

already used above (see also the remark at the end of Section 4 below).

† See, for example, Knuth, *loc. cit.*, p. 388.

THEOREM 6.2. — *Consider a squarefree polynomial F, with coefficients in the field K with q elements, whose decomposition into a product of irreducible factors is*

$$F = P_1 \cdots P_k .$$

Let L be the function defined above. The following properties hold :

(i) *If G is a polynomial whose image modulo F belongs to the nullspace of L, we have the relation*

$$F(X) = \prod_{y \in K} \text{g.c.d.}\ \big(F(X), G(X) - y\big) .$$

(ii) *If $G_1, \ldots, G_{k-1}, G_k$ is a basis of the nullspace of L, with $G_k = 1$, then, for any couple of indices i, j satisfying $1 \leq i < j \leq k$, there exists one index h, with $h < k$, and two distinct elements y_i and y_j of the field K such that the polynomial P_i divides $G_h - y_i$, whereas the polynomial P_j divides $G_h - y_j$.*

(iii) *If G_h is a polynomial defined as in* (ii), *with $h < k$, there exists one element y of field K such that the g.c.d. of the polynomials $F(X)$ and $G(X) - y$ is a nontrivial factor of the polynomial F.*

Proof

Let G be a polynomial whose image belongs to the nullspace of L. By hypothesis, the polynomial F divides $G^q - G$. But, after Theorem 6.1, we have the relation

$$G(X)^q - G(X) = \prod_{y \in K} \big(G(X) - y\big) .$$

Hence, the polynomial $F(X)$ divides the right-hand side of the formula of the theorem.

Conversely, each factor on the right-hand side divides $F(X)$, and, since these factors are two-by-two coprime, the right-hand side divides the left-hand side. Hence, we have the first assertion.

Now let $G_1, \ldots, G_{k-1}, G_k$ be a basis of the nullspace of L, with $G_k = 1$. In the isomorphism $A \sim K_1 \times \cdots K_k$ already used several times, the image of the elements $G_1, \ldots, G_{k-1}, G_k$ generates the product set $K \times \cdots \times K$. Thus, there exists some G_h sent on an element of A whose ith and jth components y_i and y_j are distinct. Obviously, we have $h < k$, since we suppose $G_k = 1$. Which demonstrates the second property.

Let G_h be a polynomial as defined in (ii) and let $(y_1, \ldots, y_k)$ be its image by the isomorphism $\operatorname{Ker} L \sim K \times \cdots \times K$. Since G_h and G_k are linearly independent, all the y_i are not equal. If these y_i have exactly m values written $z_1, \ldots, z_m$ (where we have $m \geq 2$), then, the polynomial F is written

$$(*) \qquad F(X) = \prod_{j=1}^{m} \text{g.c.d.} \left(F(X), G_h(X) - z_j\right),$$

where each of the factors on the right is nontrivial. Hence, we have the last assertion. □

3. Cost of the algorithm

Let us summarize the various steps leading to the factorization of a polynomial F with coefficients in the field $K = \mathbb{F}_q$. The degree of F is written d.

(i) We factorize F into a product of squarefree polynomials, which allows us to suppose that F is squarefree during the further steps.

(ii) We determine a basis for the nullspace of the function L defined above. We write $(G_1, \ldots, G_k)$ such a basis, where $G_k = 1$. Then k is the number of irreducible factors of F.

(iii) We apply formula (ii) of Theorem 6.2 and we first take the polynomial G_1. This theorem implies that this gives a nontrivial factorization of the polynomial F.

As long as we do not have found k distinct factors of F, we repeat the process for the successive polynomials $G_2, G_3, \ldots$ and, if necessary, up to G_{k-1}. Assertion (iii) of theorem 6.2 gives the assurance that, this way, we shall get all the irreducible factors of the polynomial F.

Now, study the cost of the algorithm. We easily verify that part (i) needs, at most, $\mathrm{O}(d^3 + d \operatorname{Log} q)$ operations in the field K.

The computation of the matrix which represents L can be done with $\mathrm{O}(d^2 \operatorname{Log} q)$ operations. Then, the computation of a basis of the nullspace of L is possible in $\mathrm{O}(d^3)$ operations. The number of necessary operations for the second step is at most $\mathrm{O}(d^3 + d^2 \operatorname{Log} q)$.

It is obvious that the computation of the g.c.d. of two polynomials P and Q, with coefficients in the field K, is possible in $\mathrm{O}(\deg P \cdot \deg Q)$ operations in K. Consequently, the third step of the algorithm needs at most $\mathrm{O}(qkd^2)$ operations in K.

In fact, the complete algorithm needs at most $\mathrm{O}(qd^3)$ operations and the last step beeing, generally, the most costly. In particular, when q is big (let us say for $q \geq 25$), this last step can become too expensive. Which is why it is useful to find a quick way to compute the points z of the field K such that they give nontrivial terms in the formula $(*)$ found during the demonstration of Theorem 6.2, that is the formula that we now write

$$(*) \qquad F(X) = \prod_{z \in S} \text{g.c.d.} \left\{F(X),\, G_h(X) - z\right\},$$

where $S = \{z_1, \ldots, z_m\}$. The next section deals with the study of the quick determination of this set S, when q is large.

4. Case where q is large

In this section, we obviously suppose F to be reducible, hence the number k of its factors is ≥ 2.

To determine the set S defined above, H. Zassenhaus¶ proposes the following method.

Set

$$H(X) = \prod_{z \in S} (X - z).$$

The relation (*) above shows that the polynomial F divides the composed polynomial $H\big(G(X)\big)$. Now we can determine H by looking for the minimal

¶ On Hensel factorization I, *J. Number Theory*, 1, 1969, p. 291–311.

polynomial of the image of G in the algebra A. Then, we get the set S by computing the roots of the polynomial H. We study the problem of computing the roots of a polynomial over a finite field in the next section.

Another method, also presented in the following section, has been proposed by Cantor and Zassenhaus and, independently, by Camion. It brings us back to the case where the z_j are 0 or 1. In other words, we determine idempotent elements of the algebra A, which necessarily belong to the nullspace of L.

Suppose the characteristic of K odd. From a polynomial $G \neq 1$ whose image modulo F belongs to the nullspace of L, we get the squareroot of an idempotent of A by taking the image modulo F of $G^{(q-1)/2}$. Indeed, if the image of the polynomial G in the product $K \times \cdots \times K$ is the element $(x_1, \ldots, x_k)$, the image of its power $G^{(q-1)/2}$ is the element $\left(x_1^{(q-1)/2}, \ldots, x_k^{(q-1)/2}\right)$ and obviously we have $x_i^{(q-1)/2} = 0$ or ± 1. If we write

$$I = \text{the set of the indices } i \text{ such that } x_i^{(q-1)/2} = 1$$

and

$$J = \text{the set of the indices } i \text{ such that } x_i^{(q-1)/2} = -1,$$

then we have the two relations

$$\text{g.c.d.}\left\{F,\, G^{(q-1)/2} - 1\right\} = \prod_{i \in I} P_i\,,$$

and

$$\text{g.c.d.}\left\{F,\, G^{(q-1)/2} + 1\right\} = \prod_{i \in J} P_i\,.$$

Hence,

$$\text{g.c.d.}\,(F,\, G) = \prod_{i \notin I \cup J} P_i\,.$$

At least one of the three previous decompositions is nontrivial, when the polynomial G satisfies the additional condition

$$G^{(q-1)/2} \not\equiv 0,\ \pm 1 \pmod{F}.$$

And the cost of the computation of these three g.c.d. [including the cost the computation of $G^{(q-1)/2}$, for which the polynomial G is supposed to be known] is at most $\mathrm{O}(d^2 \operatorname{Log} q)$ operations in the field K. The condition above occurs with a probability $> 1/2$, since, among the q^k elements G of the nullspace of the function L, there are only $2(\frac{q-1}{2})^k$ for which $G^{(q-1)/2} \equiv \pm 1 \pmod F$.

Remark

The whole principle of the algorithm of Berlekamp and of its improvements consists in determining the structure of the finite quotient algebra $A = K[X]/\bigl(F(X)\bigr)$. Supposing, as above, that F is a squarefree polynomial, whose decomposition into irreducible factors is $F = P_1 \cdots P_k$, the isomorphism of the Chinese remainder theorem is the function

$$\varphi : A \longrightarrow \prod_{i=1}^{k} K[X]/\bigl(P_i(X)\bigr),$$

defined by $\varphi(G) = (G \bmod P_1, ..., G \bmod P_k)$. In fact, we never compute the map φ — its computation is equivalent to the knowledge of the factors P_i of F — but its existence itself has been widely used.

At the beginning, the algebra A is given by its "natural" basis 1, X, $\ldots$, X^{d-1} and the computations aim mainly at finding elements G of A, whose image by φ has some remarkable properties. In the first part of the algorithm of Berlekamp, the polynomials G must be sent on a basis of the subspace $K \times \cdots \times K$ of the product $K_1 \times \cdots \times K_k$, where K_i designates the field $K[X]/\bigl(P_i(X)\bigr)$. In the final step, we try to determine the elements y_i of K, such that, for a given polynomial G which belongs to the nullspace of L, we have

$$\varphi(G) = (y_1, \ldots, y_k) \in K \times \cdots \times K.$$

The method of Camion-Cantor-Zassenhaus determines polynomials, such that the previous y_i are equal to 0 or ± 1. Finally, the search for the primitive idempotents of the algebra A is equivalent to the determination of polynomials whose image by φ is the canonical basis of the K–vector space $K \times \cdots \times K$ (an element of a ring e is called an *idempotent* when it is nonzero and when it satisfies $e \cdot e = e$; an idempotent is said to be *primitive* when — for any other idempotent e' — we have the relation $e \cdot e' = 0$).

5. Decomposition into a product of irreducible factors of given degree

Let h be an integer, $h \geq 2$. The polynomial

$$R_h(X) = \text{g.c.d.}\left\{F(X),\, X^{q^h} - X\right\}$$

gives the product of the irreducible factors of F whose degree divides h. To compute this polynomial, we, first, compute

$$X^{q^h} \quad \left(\text{mod} F(X)\right),$$

which is possible in $\text{O}(h\, d \text{ Log } q)$ operations, then, we can compute the g.c.d. R_h in $\text{O}(d^2)$ operations.

Moreover, if the first computation has been done for the positive rational integer h, we easily notice that the same computation for the next rational integer $h + 1$ is possible in $\text{O}(d \text{ Log } q)$ operations. For h varying from 1 to H, we get $\text{O}(d\, H \text{ Log } q + d^2\, H)$ operations.

We also note that the computation of the polynomials R_h, for the indices $1 \leq h \leq H$, allows us to easily obtain the product S_h of the irreducible factors of F whose degree is equal to h (for h fixed, with $1 \leq h \leq H$). The polynomial S_h is given by the formula

$$S_h = \frac{R_h}{\prod\limits_{j|h\,;\, j<h} S_j},$$

or by

$$S_h = \frac{R_h}{\text{g.c.d.}\left(R_h,\ \prod\limits_{j|h\,;\, j<h} R_j\right)}.$$

6. *The method of MacEleice*†

First, demonstrate a lemma, then a proposition, on which this algorithm is based.

LEMMA 2. — *Let F and G be two squarefree polynomials, with coefficients in the field $\mathbb{F}_q$, such that F divides G. We put*

$$A_F = \mathbb{F}_q[X]/(F) \quad \text{and} \quad A_G = \mathbb{F}_q[X]/(G)$$

and we respectively write L_F and L_G for the corresponding functions L (defined in Section 4.1). Also, let $\pi : A_G \longrightarrow A_F$, be the canonical surjection. Then, we have

$$\text{Ker } L_F = \pi(\text{Ker } L_G).$$

Proof

Let $P_1, \ldots, P_r$ be the irreducible divisors of the polynomial G, in such an order that we have $F = P_1 \cdots P_k$. Then the primitive idempotents e_1, $\ldots$, e_r of A_G are the images modulo G of polynomials E_i that satisfy

$$E_i \equiv \delta_{i\,j} \pmod{P_j}, \qquad j = 1, \ldots, r.$$

It is obvious that the elements $\pi(E_1)$, $\ldots$, $\pi(E_k)$ are the primitive idempotents of the algebra A_F. Hence, we have the conclusion. □

PROPOSITION 6.8. — *Let F be a polynomial over the finite field $\mathbb{F}_q$ that satisfies $F(0) \neq 0$. Let t be the period of F (that is, the smallest positive integer j such that F divides $X^j - 1$).*

Then, the polynomials $T_i = T_{i,F}$ defined, for the indices $i = 1$, $\ldots$, $t-1$, by the formulas

$$T_i(X) = X^i + X^{iq} + X^{iq^2} + \cdots + X^{iq_i} \bmod F(X),$$

where the rational integer q_i is the smallest power of q, such that the polynomial F divides $X^i - X^{iq_i}$, generate the nullspace of the mapping L computed during the algorithm of Berlekamp.

---

† The reference for the algorithm of R. J. MacEleice which we study in this section is "Factorization of polynomials over finite fields", *Math. of Computation*, 23 (108), 1969, p. 861–868.

Proof

By definition, it is obvious that each of the T_i belongs to the nullspace of the function L. First, suppose $F = X^t - 1$, and put $H = X^t - 1$. Also, put $q_i = q^{m_i - 1}$; then m_i is the smallest positive integer such that t divides the product $i\,q^{m_i-1}$. The exponents of $T_{i,\,H}$ are the various residues modulo t of the m_i numbers $i,\ i\,q,\ i\,q^2,\ \ldots i\,q^{m_i-1}$.

Consider, now, a polynomial G belonging to the nullspace of L_H. Then, we have the relation

$$G^q \equiv G \pmod{X^t - 1},$$

and if

$$G(X) = a_0 + a_1 X + \cdots + a_{t-1} X^{t-1},$$

we deduce the relation $a_n = a_{n'}$ when there exists a rational integer u such that we have

$$n \equiv n'\,q^u \pmod{t},$$

which shows that G is a linear combination of the polynomials $T_{i,\,H}$.

Now, let F be a strict divisor of H. After Lemma 6.2 and the demonstration above, the family of the polynomials $T_{i,\,H} \bmod F$, for $i = 1, \ldots, t-1$, generates the nullspace of the function L_F. But it is easy to see that each polynomial $T_{i,\,H}$ is equal to $T_{i,\,H} = aT_{i,\,F}$, for some integer a dependent on i. Hence, we have the conclusion. □

This result is mainly useful when t is small. It is the case when all the irreducible factors of F are of the same degree, and we can carry out this condition using the algorithm of the previous section.

Remark

Suppose that the polynomial F is the product of k irreducible distinct factors, of the same degree d', and let t be the period of F. Then, the period t divides $q^{d'-1}$. Let t' be a divisor of t such that the quotient t/t' is prime. Then, there exists, at least, one irreducible factor of F whose period does not divide t'. Hence — unless all the irreducible factors of F have a period equal to t — the polynomial

$$\text{g.c.d.}\left\{F(X),\ X^{t'}-1\right\}$$

is a nontrivial factor of F for at least one of the possible values t'.

5. Search for the roots of a polynomial in a finite field

1. A first reduction

As we have seen, the computation of the g.c.d. of the polynomials $F(X)$ and $X^q - X$ allows us to search only for the roots of a polynomial which splits completely in the field $\mathbb{F}_q$. Moreover, we suppose that the polynomial F has a nonzero constant term.

We can also notice that the procedures laid out by Berlekamp* allow us to study only the case where F belong to $\mathbb{F}_p[X]$, and where F splits completely in this ring.

2. Case of the polynomials which split completely in $\mathbb{F}_q$

In this section we assume that q is odd. Consider a polynomial F such that

$$F(X) = \prod_{i=1}^{d}(X - s_i),$$

where the s_i are nonzero (unknown) elements of the field $\mathbb{F}_q$. The method for factorizing into a product of squarefree polynomials allows us to study only the case where the s_i are distinct.

As we know, we can write

$$F(X) = \text{g.c.d.}\left\{F(X),\ X^{(q-1)/2}-1\right\}\cdot\text{g.c.d.}\left\{F(X),\ X^{(q-1)/2}+1\right\}.$$

This formula separates the roots s_i in two sets : the roots that are squares in the multiplicative group $\mathbb{F}_q^*$, and the roots that are not squares in this

* Factoring polynomials over large finite fields, *Math. of Computation*, 24 (111), 1970, p. 713–735, Sect. 5 and 6.

group. If this decomposition is trivial, we replace the polynomial $F(X)$ by some translated polynomial, say $F(X-a)$. We admit that the probability for a value of a to lead to a trivial decomposition is close to 2^{-d+1}. This implies, that, in general, we should reach a nontrivial decomposition of F after very few choices of the value a of the translation.

Example

Consider the polynomial

$$F(X) = X^4 - 3X^3 - 3X^2 - 3X + 1$$

over the field $\mathbb{F}_7$.

We have

$$\begin{aligned}\text{g.c.d.}\left\{F(X),\, X^3+1\right\} &= \text{g.c.d.}\,(-3X^3 - 3X^2 + 3X + 1,\, X^3 + 1) \\ &= \text{g.c.d.}\,(-3X^2 + 3X - 3, X^3 + 1) = X^2 - X + 1,\end{aligned}$$

and

$$F(X) = (X^2 - X + 1)\,(X^2 - 2X + 1)\,.$$

The second of these factors is

$$(X^2 - 2X + 1) = (X-1)^2\,.$$

Notice that the first factor, $X^2 - X + 1$, has a discriminant equal to -3. The congruence $-3 \equiv 2^2 \pmod 7$ and the well-known formulas to solve a second-degree equation, give

$$X^2 - X + 1 \equiv (X-3)(X-5) \pmod 7,$$

whence, we obtain the complete decomposition into linear factors

$$F(X) \equiv (X-1)^2\,(X-3)\,(X+2) \pmod 7.$$

3. Solution of the binomial equation $X^d = a$

Consider an equation

$$(\mathcal{E}) \qquad x^d = a,$$

where a is a nonzero element of K. We try to determine all the solutions of equation $(\mathcal{E})$ that belong to the field K. Let h be the order of the element a in the group K^* of the nonzero elements of K. Put $k = (q-1)/h$. The group K^* is cyclic and admits a generator z, such that we have $a = z^k$. Put $x = z^u$, where u is unknown. The equation $(\mathcal{E})$ is equivalent to the congruence

$$ud \equiv k \pmod{q-1}.$$

Let e be the g.c.d. of the integers d and $q-1$. Put $d = ed'$ and $q-1 = eq'$. For the equation $(\mathcal{E})$ to have a solution in K, e must divide k.

Conversely, if $k = ek'$, where k' is an integer, then the previous congruence is equivalent to

$$ud' \equiv k' \pmod{q'}$$

and the equation $\mathcal{E}$ has e solutions in K.

Notice, finally, that the condition "e divides k" is equivalent to the relation $a^{d'} = 1$. The equality $a^{d'} = 1$ can be tested in, at most, $\mathrm{O}(\mathrm{Log}\, q)$ multiplications in the field K. From now on, we shall always assume that a satisfies the condition $a^{d'} = 1$.

To solve equation $(\mathcal{E})$, we can first solve the equation $y^e = a$, then the equation $x^{d'} = y$. This reduces the general problem to the two following particular cases :

(i) $(d, q-1) = 1$,

(ii) d divides $q-1$.

The first case is very easy to solve : let u and v be two integers such that we have

$$ud + v(q-1) = 1, \qquad \text{with } 1 \le u < q;$$

then $x = a^u$ is a solution to $(\mathcal{E})$ (indeed $(a^u)^d = a^{1-v(q-1)} = a$), and, moreover, $(\mathcal{E})$ has no other solution in the field K. Also note that the computation of the rational integer u and of a^u can both be done in at most $\mathrm{O}(\mathrm{Log}\, q)$ operations.

Example

If q is congruent to -1 modulo 6, the equation $X^3 = a$ is solvable in the field $\mathbb{F}_q$ in $\mathrm{O}(\mathrm{Log}\, q)$ operations. More generally, we can find a solution in $\mathbb{F}_q$ of any cubic equation in $\mathrm{O}(\mathrm{Log}\, q)$ operations, at least when we know one element of K^* that is not a square (use the formulas of Cardan and the solution of the equation $X^2 = b$ given above).

Now, let us study now case (ii). First, consider the simplest case of the equation $x^2 = a$, with q odd and $a^{(q-1)/2} = 1$. Our presentationfollows that of D.N. Lehmer†. We extend this work to the case where q is not a prime number, and we bring some simplifications.

If q is equal to $4k+3$, the solutions are $x = \pm a^{k+1}$. Indeed, we have the relation $x^2 = a^{2k+2} = a^{1+(q-1)/2} = a$. Hence, a computation of the solutions in $\mathrm{O}(\mathrm{Log}\, q)$ operations. When q is equal to $4k+1$, the following proposition gives a way to compute some solutions.

PROPOSITION 6.9. — *Let q be a power of an odd prime number, say $q = 4k+1$. Let a and b be two integers such that a is a square in the multiplicative group $\mathbb{F}_q^*$ but $b^2 - 4$ is not one. And let v_n be the linear recursive sequence defined by the conditions*

$$v_0 = 2, \quad v_1 = b, \quad v_n = bv_{n-1} - av_{n-2}, \quad \text{for} \quad n \geq 2.$$

Then the solutions of the equation $x^2 = a$ in $\mathbb{F}_q$ are $x = \pm(\frac{1}{2})\, v_{(q+1)/2}$.

Proof

Let K' be the field of cardinality q^2 and let α and β be the roots in the field K' of the equation $X^2 - bX + a = 0$. We can easily verify that we

† Computer Technology Applied to the Theory of Numbers, *Studies in Number Theory*, p. 117–151, ed. W.J. Leveque, Englewood Cliff, Prentice Hall, 1969.

have $v_n = \alpha^n + \beta^n$ for all $n \geq 0$. Hence the relation

$$\begin{aligned} v^2_{(q+1)/2} &= \left\{ \alpha^{(q+1)/2} + \beta^{(q+1)/2} \right\}^2 \\ &= \alpha^{q+1} + \beta^{q+1} + 2\,(\alpha\beta)^{\,(q+1)/2}. \end{aligned}$$

The transformation $y \longmapsto y^q$ leaves the above equation invariant, hence, the number α^q is one of its roots. However $\alpha^q \notin \mathbb{F}_q$ (since $b^2 - 4$ a is not a square in $\mathbb{F}_q$), and hence $\alpha^q = \beta$. For the same reason, $\beta^q = \alpha$. Moreover, since a is a nonzero square of $\mathbb{F}_q$, we have the relation $a^{(q-1)/2} = 1$. In fact,

$$v^2_{(q+1)/2} = 2\,\alpha\beta + 2\,a^{(q+1)/2} = 4a\,.$$

Hence, we have the result. □

We can compute $v_{(q+1)/2}$ in $\mathrm{O}(\mathrm{Log}\, q)$ operations using any one of the formulas (easy to verify)

$$v_{2m} = v_m^2 - 2\,a^m \qquad \text{or} \qquad \begin{pmatrix} v_{m+1} \\ v_m \end{pmatrix} = \begin{pmatrix} b & -a \\ 1 & 0 \end{pmatrix}^m \begin{pmatrix} v_1 \\ v_0 \end{pmatrix}.$$

Hence, we have shown that if a second-degree equation on the field $\mathbb{F}_q$ is solvable in this field, then we can find its solutions in at most $\mathrm{O}(\mathrm{Log}\, q)$ operations — provided we know one element of this field which is not a square.

Now, let us look at the general case of condition (ii), using a procedure inspired by Gauss‡.

Suppose that the exponent d divides $q - 1$ and that a satisfies the condition $a^{(q-1)/d} = 1$. Also, assume that d is prime, which does not restrict the generality (since any extraction of a root is a sequence of extractions of prime order roots). Now, put $q - 1 = d^r q'$ and $k = d^s k'$, where d divides neither q' nor k'. Suppose that we know one element y of K^* of order d^r. Let w and t be two integers, such that we have $wd = 1 + tq'$. If x is a solution of equation $(\mathcal{E})$, then we have the relation

$$(x\,a^{-w})^d = a^{-tq'}$$

‡ *Disquisitiones arithmeticae*, Fleischer, Leipzig, 1801, Sec. 67–68.

in which the right-hand side has an order which divides d^r. The solutions x of equation $(\mathcal{E})$ are given by the formulas

$$x = y^j\, a^w, \quad \text{with} \quad wd \equiv 1 \pmod{q'}.$$

Put $wd = 1 + tq'$. We have $y = z^{q'm}$, where d does not divide the rational integer m, and where z is the generator of K^* chosen above. We, now, have to find an integer j such that we have

$$mj + td^{s-1}k' \equiv 0 \pmod{d^{r-1}}.$$

Hence, at most d^{r-1} tries of this rational integer j are needed. Finally, if x is a solution in K, all the other solutions x' of equation $(\mathcal{E})$ that belong to this field are given by

$$x' = y^{i\, d^{r-1}} x, \qquad i = 0, 1, \ldots, d-1.$$

We can now find all the solutions of equation $(\mathcal{E})$ in at most $\mathrm{O}(d^r \operatorname{Log} q)$ operations in the field K, provided we know some element of the multiplicative group K^* which is of order d^r.

Let us summarize this study.

PROPOSITION 1.10. — *Let d be a prime number that divides $q-1$. We suppose that we know one element of the field $\mathbb{F}_q$ of order d^r, where d^r is the highest power of d that divides $q-1$. Let a be any nonzero element of the field $\mathbb{F}_q$ that satisfies $a^{(q-1)/d} = 1$. Then, we can compute all the solutions of the equation $x^d = a$, for x belonging to $\mathbb{F}_q$, in $\mathrm{O}(d^r \operatorname{Log} q)$ operations in the finite field $\mathbb{F}_q$.*

Exercises

1. Some lower bound for $\varphi(n)$

1°) Demonstrate that the sequence

$$\sum_{n\geq 2}\Big\{\mathrm{Log}\Big(\frac{1}{1-1/n}\Big)-\frac{1}{n}\Big\}$$

is convergent.

2°) Let $(c_n)_{n\geq 1}$ be a sequence of real numbers. We put

$$C(t)=\sum_{n\leq t} c_n\,.$$

Let $f:\big[1,\infty\big]\longrightarrow R$ be a function. Demonstrate the relation

$$\sum_{n\leq x} c_n\, f(n)=\sum_{n\leq x-1} C(n)\,\big\{f(n)-f(n-1)\big\}+C(x)\,f([x])\,.$$

Deduce that, in the case where f is continuously derivable, we have

$$\sum_{n\leq x} c_n\, f(n)=C(x)\,f([x])-\int_0^x C(t)\,f'(t)\,dt\,.$$

3°) Demonstrate that, if p is a prime number and if p^h divides $n\,!$, then we have the inequality $h\geq \frac{n}{p}-1$.

Deduce the upper bound

$$\sum_{p\leq x}\frac{\mathrm{Log}\,p}{p}\leq \mathrm{Log}\,x+\mathrm{O}(1)\,.$$

4°) Demonstrate the inequality

$$\sum_{p|n}\frac{1}{p}\leq 1+\int_1^x\frac{\mathrm{Log}\,(t)+c}{t(\mathrm{Log}\,t)^2}\,dt\,,$$

where c is a constant.

5°) Deduce the lower bound

$$\frac{\varphi(n)}{n} \geq \frac{c'}{\operatorname{Log}\operatorname{Log} n},$$

where c' is a constant.

2. Solution of the equation $x^p = a$ in a finite group

Let G be a finite group of cardinality n and let p be a prime number that divides n, so that $n = p^r n'$, where p does not divide n'.

1°) We put $\mathrm{G}(p) = \{x \in G\,;\ \exists k,\ \mathrm{order}(x) \mid p^k\}$. Demonstrate that

$$\mathrm{G}(p) = \{x \in G\,;\ x^{p^r} = 1\},$$

where 1 is the neutral element of G.

2°) Let a and x be two elements of G such that $x^p = a$.

(i) Demonstrate that $a^{n/p} = 1$.

(ii) Let u be an integer such that $u^p \equiv 1 \pmod{n'}$. Demonstrate that $x\,a^{-u}$ belongs to $\mathrm{G}(p)$.

(iii) Deduce that knowing $\mathrm{G}(p)$ allows to get a procedure for the solution of the equation $x^p = a$ in the group G, for a fixed, in a number of operations in G at most $\mathrm{O}(p^r \operatorname{Log} n)$.

(iv) In the particular case where a satisfies the relation $a^{n'} = 1$, demonstrate that the element $x = a^u$ is a solution of the equation $x^p = a$.

3. Statistics in $\mathbb{F}_q[X]$

To a subset E of $\mathbb{F}_q[X]$, we associate a formal serie $S(t)$, called the *generating function* of E, defined by

$$S(t) = S_E(t) = \sum_{m=0}^{\infty} a_m\, t^m,$$

where $a_m = \mathrm{Card}\{F \in E\,;\ \deg F = m\}$.

1°) Demonstrate the following properties :

(i) If $E = \{F \in \mathbb{F}_q[X]\,;\, F \text{ monic}\}$ then

$$S(t) = S_E(t) = \frac{1}{1-qt}\cdot$$

(ii) If E is the disjoint union of two sets E' and E'' then

$$S_E(t) = S_{E'}(t) + S_{E''}(t)\,.$$

2°) Let E_1, E_2, …,E_n be sets of monic polynomials such that

$$(F \in E_i,\, G \in E_j,\, i \neq j) \implies \text{g.c.d.}\,(F, G) = 1\,.$$

We call E the set of the polynomials which are equal to $F_1 \cdots F_n$, where F_i runs over E_i for any i. Demonstrate that

$$S_E(t) = \prod_{i=1}^{n} S_i(t)\,,$$

where S_i is the generating function of E_i for $i = 1, 2, \ldots, n$. Generalize this result to the case of an infinite sequence $(E_i)_{i\geq 0}$ of parts of $\mathbb{F}_q[X]$ having some suitable properties.

3°) For $m = 1, 2, \ldots$, let R_m be a set of i_m irreducible monic polynomials of degree m. Let E be the set of the monic polynomials whose all irreducible factors belong to the union R of the R_m. Demonstrate the relation

$$S_E(t) = \prod_{m=1}^{\infty} (1 - t^m)^{-i_m}\,.$$

Deduce the formula

$$\frac{1}{1-qt} = \prod_{m=1}^{\infty} \left(\frac{1}{1-t^m}\right)^{I_m},$$

where I_m is the number of polynomials of $\mathbb{F}_q[X]$, irreducible, monic, and of degree m.

Let E' be the set of the monic polynomials that are products of irreducible factors, all distinct, and belonging to R. Demonstrate that

$$S_{E'}(t) = \prod_{m=1}^{\infty} \left(1 - t^m\right)^{i_m}.$$

4°) Demonstrate that the generating function of the set of squares of the monic polynomials is

$$\frac{1}{1 - qt^2}.$$

5°) § Let Q be the set of the monic and squarefree polynomials of $\mathbb{F}_q[X]$, and let

$$Q' = \{F \in Q\,;\, \omega(Q) \text{ even}\} \quad \text{and} \quad Q'' = \{F \in Q\,;\, \omega(Q) \text{ odd}\}.$$

Demonstrate the following relations

$$S_Q(t) = \prod_{m=1}^{\infty} (1 + t^m)^{I_m} = \frac{1 - q\,t^2}{1 - qt},$$

$$S_{Q'}(t) - S_{Q''}(t) = \prod_{m=1}^{\infty} (1 - t^m)^{I_m} = 1 - qt.$$

Deduce that, for $m \geq 2$, the number of squarefree polynomials of degree m having an even number of irreducible factors and the number of those having an odd number of irreducible factors are equal.

Demonstrate the formulas

$$S_{Q'}(t) = \frac{1 - qt + (1 - qt^2)/(1 - qt)}{2},$$

and

$$S_{Q''}(t) = \frac{-1 + qt + (1 - qt^2)/(1 - qt)}{2}.$$

§ Reference : E. R. Berlekamp, *Algebraic Coding Theory*, § 3.3.

6°) Demonstrate that the generating function of the set of the monic polynomials that have no root in the field $\mathbb{F}_q$ is

$$\frac{(1-t)^q}{1-qt} = \prod_{m=2}^{\infty} (1-t^m)^{I_m}.$$

Generalize this result to the set of the monic polynomials that have no irreducible factor of degree smaller than or equal to k, where k is a fixed positive integer.

4. Cyclotomic polynomials

We call Φ_n the cyclotomic polynomial of order n, and, since it is more convenient here, we also put $F(n\,;\,X) = \Phi_n(X)$.

1°) Demonstrate that, if p is a prime number that does not divide the integer m, then we have the formula

$$F(m\,p^k\,;\,X) = F(m\,p\,;\,X^{p^{k-1}}),$$

for any positive integer k. Under the same hypothesis, demonstrate that

$$F(m\,p\,;\,X) = \frac{F(m\,;\,X^p)}{F(m\,;\,X)}.$$

Then, show that, for any odd integer n, $n \geq 3$, we have the relation

$$F(2\,n\,;\,X) = F(n\,;\,X).$$

2°) Deduce the relations

$$\Phi_n(1) = \begin{cases} 0 & \text{if} \quad n = 1, \\ p & \text{if} \quad n = p^k, \;\; k \geq 1, \\ 1 & \text{if} \quad \omega(n) \geq 2, \end{cases}$$

and

$$\Phi_n(-1) = \begin{cases} 0 & \text{if} \quad n = 2, \\ -2 & \text{if} \quad n = 1, \\ 2 & \text{if} \quad n = 2^k, \;\; k \geq 2, \\ p & \text{if} \quad n = 2\,n', \;\; n' = p^k, \;\; k \geq 1, \\ 1 & \text{otherwise.} \end{cases}$$

3°) Let n be a positive integer ≥ 2 and let d be some divisor of n, with $1 \leq d < n$. Demonstrate that the polynomial $\Phi_n(X)$ divides the polynomial $(X^n - 1)/(X^d - 1)$.

4°) Suppose that the integers n and q are relatively prime. Demonstrate that the cyclotomic polynomial Φ_n can, then, be decomposed in the field $\mathbb{F}_q$ into $\varphi(n)/d$ irreducible factors, each of degree d, where d is the smallest positive integer, such that $q^d \equiv 1 \pmod{n}$.

5. Congruences of Lucas

Let p be a prime number. The following congruences are all modulo p. They were demonstrated, for the first time, by E. Lucas.

1°) Let a be an integer, with $0 \leq a < p$. Demonstrate the congruences

$$(1+X)^{np+a} \equiv (1+X^p)^n \, (1+X)^a.$$

2°) Let b be an integer, with $0 \leq b < p$. Considering the coefficients of the term X^{kp+b} in the previous formula, demonstrate the relation

$$\binom{np+a}{kp+a} = \binom{n}{k}\binom{a}{b}.$$

Deduce the fact that the binomial coefficient $\binom{np+a}{kp+b}$ is divisible by p, when a and b satisfy $a < b$.

3°) Show, as a conclusion, that if the decomposition of the rational integer n in scale p is $n = (n_j \ldots n_1)$ and that the one of k is given by $k = (k_j \ldots k_1)$ (where the k_i are nonnegative) then we have

$$\binom{n}{k} = \binom{n_1}{k_1} \cdots \binom{n_j}{k_j}.$$

6. Let L be the finite field with q^n elements. Demonstrate that the number of $n$tuples of elements of L that constitute a basis of the field L over the field $\mathbb{F}_q$ is equal to

$$(q^n - 1)\,(q^n - q) \cdots (q^n - q^{n-1}).$$

7. Extensions of the field $\mathbb{F}_q$

For simplification, in this exercise, the field $\mathbb{F}_q$ will be written K. We consider a fixed positive integer n, $n \geq 2$, and we call L the finite field with q^n elements. We also take a polynomial $f \in K[X]$, irreducible over K and of degree equal to n.

1°) Demonstrate that the polynomial f is squarefree.

2°) Demonstrate that there exists at least one element α of L such that $f(\alpha) = 0$.

Also, demonstrate that, for such an α, we have $L = K[\alpha]$.

3°) Demonstrate that f can be decomposed into linear factors in the field L and that its roots are n distinct elements of L :

$$\alpha,\ \alpha^q,\ \alpha^{q^2},\ \ldots, \alpha^{q^{n-1}}.$$

4°) If a and b are two roots of f in L, demonstrate the equality

$$\operatorname{order}(a\,;\ L^*) = \operatorname{order}(b\,;\ L^*).$$

5°) Demonstrate that the set G of the automorphisms $\sigma : L \longrightarrow L$ that satisfy the condition $\sigma(x) = x$ for any element x of the field K constitute a group. Let j be an integer, with $0 \leq j < n$. Demonstrate that there exists only one K–linear function φ of the set L into itself, that satisfies

$$\varphi(\alpha) = \alpha^{q^j}, \quad \text{and} \quad \varphi(\alpha^i) = \varphi(\alpha)^i, \quad \text{for all} \quad i \geq 0.$$

This function will be written σ_j. Demonstrate that σ_j belongs to G, since

$$G = \{\sigma_0, \sigma_1, \ldots, \sigma_{n-1}\}.$$

Deduce that G is a cyclic group isomorphic to the additive group $\mathbb{Z}/n\mathbb{Z}$.

6°) Let Tr : $L \longrightarrow L$ the K-linear function defined by the relations

$$\operatorname{Tr}(\alpha^i) = \alpha^i + \alpha^{iq} + \cdots + \alpha^{iq^{n-1}}$$

for the values $i = 0, 1, \ldots, n$. Demonstrate that $\operatorname{Tr}(L) = K$ and that, for any element β of L, we have

$$\operatorname{Tr}(\beta^p) = \operatorname{Tr}(\beta)^p \quad \text{and} \quad \operatorname{Tr}(\beta^q) = \operatorname{Tr}(\beta).$$

If β belongs to L, demonstrate that the function T_β of L into itself, defined for any x in L by the formula

$$T_\beta(x) = \operatorname{Tr}(\beta \cdot x),$$

is a K-linear function whose image is contained in the field K.

If β and γ are two distinct elements of L, demonstrate that the two functions T_β and T_γ are different.

Deduce that, if $T : L \longrightarrow K$ is a K–linear function, then there exists β in the field L, such that $T = T_\beta$.

Let $a_1, \ldots, a_n$ be a basis of L as a K–vectorial space. Prove that there exists a basis $b_1, \ldots, b_n$ of L, such that

$$\operatorname{Tr}(\alpha_i \cdot \beta_j) = \delta_{ij},$$

where as usual δ_{ij} is the symbol of Kronecker.

7°) For β in L, we define the vector $V(\beta) = (\beta_1, \ldots, \beta_n)$, where

$$\beta_i = \beta^{q^{i-1}} \quad \text{for} \quad i = 1, 2, \ldots, n.$$

Demonstrate that $a_1, \ldots, a_n$ is a basis of L (as a K–vectorial space) if, and only if, the vectors $V(a_1), \ldots, V(a_n)$ are K–linearly independent. Deduce that $\alpha_1, \alpha_2, \ldots, \alpha_n$ is a basis of L if, and only if, the matrix A that is defined by

$$A = (\alpha_{ij}), \quad \text{where} \quad \alpha_{ij} = {\alpha_i}^{q^{j-1}}, \text{ for } 1 \leq i, j \leq n.$$

has a nonzero determinant.

8°) If $a_1, \ldots, a_n$ are elements of L, we put

$$b_{ij} = \mathrm{Tr}(\alpha_i \cdot \alpha_j), \qquad \text{for } 1 \leq i,\, j \leq n\,;$$

and we call B the matrix of the coefficients b_{ij}. Demonstrate the relation

$$B = {}^tA \cdot A \quad (\text{where } {}^tA \text{ is the transposed matrix of } A).$$

Deduce the following criterion

$$\alpha_1, \ldots, \alpha_n \text{ is a basis of } L \iff \det(B) \neq 0.$$

8. Companion matrix

Let K be a field and let f be a monic polynomial with coefficients in this field, that is given by the formula

$$f(X) = X^n - a_{n-1}\,X^{n-1} - a_{n-2}\,X^{n-2} - \cdots - a_0\,.$$

Consider the matrix $A = A_f$ defined by

$$A = \begin{pmatrix} 0 & 0 & \ldots & 0 & a_0 \\ 1 & 0 & \ldots & 0 & a_1 \\ 0 & 1 & \ldots & 0 & a_2 \\ \ldots & & \ldots & & \ldots \\ 0 & 0 & \ldots & 1 & a_{n-1} \end{pmatrix}.$$

1°) We call e_i, $1 \leq i \leq n$, the column vector of components (δ_{ij})¶, where $1 \leq j \leq n$. Demonstrate the formulas

$$A(e_i) = e_{i+1}\,, \qquad \text{for } 1 \leq i < n,$$

and

$$A(e_n) = a_0 e_1 + a_1 e_2 + \cdots + a_{n-1} e_n\,.$$

Demonstrate that the n matrices I, A, ... and A^{n-1} are linearly independent over the field K and that we have $F(e_1) = 0$ where

$$F = f(A) = A^n - a_{n-1}\,A^{n-1} - a_{n-2}\,A^{n-2} - \cdots - a_0\,I\,.$$

Deduce that the matrix F is zero and that f is the minimal polynomial of the matrix A.

¶ As usual, (δ_{ij}) represents Kronecker's symbol.

2°) Let $K[A]$ be the set of the matrices that are linear combinations with coefficients in K of the powers of the matrix A and of the identity matrix I. Demonstrate that the function

$$\varphi : K[X]/(f(X)) \longrightarrow K[A],$$

that, for $i = 1, 2, \ldots, n-1$, associates A^i to X^i and which sends 1 on the matrix I, is an isomorphism of rings and of K–vector spaces. Deduce, in particular, that if K is a finite field and L is a finite extension of K then, we can represent L as a ring of matrices over K. Give such an example for $K = \mathbb{F}_3$ and $L = \mathbb{F}_9$.

3°) Demonstrate that, for any $k \geq 1$, we have the equivalence

$$f(X) \text{ divides } X^k - 1 \iff A^k = I.$$

9. Splitting field

Let K be the field $\mathbb{F}_q$ and let f be a polynomial, with coefficients in K, and of degree ≥ 2. We call L the smallest extension of the field K in which the polynomial f can be decomposed into a product of first-degree polynomials (such an extension exists, see Exercise 38 of Chapter 3). We call $r = r(f)$ the degree of L as a vector space over the field K.

1°) Let $f_1, \ldots, f_k$ be the monic polynomials, irreducible over K, that divide f. Demonstrate the relation

$$r(f) = \text{l.c.m.}(d_1, \ldots, d_k) \quad \text{where} \quad d_i = \deg(f_i), \quad \text{for } i = 1, \ldots, k.$$

2°) Demonstrate that we have

$$\max\{\, r(f)\,;\ f \in K[X],\ \deg(f) = n \,\} = g(n)$$

where $g(n)$ is the function defined by

$$g(n) = \max\{\, \text{l.c.m.}(m_1, \ldots, m_s)\,;\ s,\ m_1, \ldots, m_s,\ m_1 + \cdots + m_s \leq n \,\}.$$

Demonstrate the estimate†

$$\operatorname{Log} g(n) \approx \sqrt{n \operatorname{Log}(n)}\,.$$

3°) Let, as in the first question, k be the number of irreducible factors of the polynomial f. Demonstrate the inequality

$$r(f) \leq \binom{n}{k}^{k}.$$

Deduce that, for any fixed positive number ε and for any integer n big enough, we have "in general"

$$r(f) \leq \exp\bigl\{(1+\varepsilon)\,(\operatorname{Log} n)^2\bigr\}.$$

[Remark : we can show that the inequality $r(f) \geq \exp\{(1-\varepsilon)\,(\operatorname{Log} n)^2\}$ is also true "in general" ; but it is much more difficult to prove it. ‡

10. Period of a polynomial

Let K be "the" finite field with q elements. Consider a nonconstant univariate polynomial f of degree equal to n, with coefficients in K, that satisfies the condition $f(0) \neq 0$.

1°) Demonstrate that there exists a positive integer k such that the polynomial $f(X)$ divides $X^k - 1$. The smallest such integer is called the *period* of f, and is denoted $\operatorname{per}(f)$. Demonstrate the inequality

$$\operatorname{per}(f) \leq q^n - 1\,.$$

2°) If $f = gh$, where the polynomials f and g are relatively prime, demonstrate the relation

$$\operatorname{per}(f) = \text{g.c.d.}\bigl\{\operatorname{per}(g),\, \operatorname{per}(h)\bigr\}.$$

† See the article of J.L. Nicolas . — Ordre maximal d'un element du groupe S_n des permutations et "highly composite numbers", *Bull. Soc. Math. France*, 97 (108), 1969, p. 861–868.

‡ See the article of M. Mignotte and J. L. Nicolas . — Statistiques sur $\mathbb{F}_q[X]$, *Ann. Inst. Henri Poincaré*, 19 (2), 1983, p. 113–121.

3°) Demonstrate that if f is irreducible, then the period of the polynomial f divides $q^n - 1$.

4°) Moreover, we suppose that f is monic and can be decomposed as

$$f = f_1^{n_1} \cdots f_k^{n_k},$$

where the f_i are distinct, irreducible, and monic.

Demonstrate the relation

$$\operatorname{per}(f) = e\,p^s \quad \text{where} \quad e = \text{l.c.m.}\left\{\operatorname{per}(f_1), \dots, \operatorname{per}(f_k)\right\}$$

and where the rational integer s is defined by the inequalities

$$p^s \geq \max\{n_1, \dots, n_k\} > p^{s-1}.$$

5°) Compare $\operatorname{per}\{f(X)\}$ and $\operatorname{per}\{f(-X)\}$.§

6°) If f is monic, let r be the smallest integer such that X^r is congruent to an element of the field K modulo the polynomial f and take a such that $X^r \equiv a$. Demonstrate that we have

$$\operatorname{per}(f) = rh,$$

where h is the order of a in the multiplicative group K^*.

11. Demonstrate that the product of the monic irreducible polynomials of degree n over the field $\mathbb{F}_q$ is given by the formula

$$\prod_{d|n} \left(X^{q^d} - X\right)^{\mu(n/d)}.$$

§ See R. Lidl and H. Niederreiter . — *Finite Fields*, Addison Wesley, Reading (Mass.),1983.

12. Let K and L be the finite fields having, respectively, q and $q' = q^m$ elements. Let

$$f = \sum_{i=0}^{q'-1} a_i X^i$$

be a polynomial with coefficients in L.

Demonstrate that the following properties are equivalent :

(i) $f(L) \subset K$,

(ii) $X^{q'} - X$ divides $f(X)^q - f(X)$,

(iii) we have $a_{q'-1} = a_{q'-1}^q$ and $j \equiv qi \Longrightarrow a_j = a_i^q$ for the integers $0 \leq i, j < q - 1$.

13. Let $f : \mathbb{F}_q \longrightarrow \mathbb{F}_q$ be any function. Demonstrate that there exists a polynomial and only one $F \in \mathbb{F}_q[X]$, of degree $< q$, and such that we have the equality $f(x) = F(x)$ for every element x of $\mathbb{F}_q$. Demonstrate the formula

$$F(X) = \sum_{\alpha \in \mathbb{F}_q} f(\alpha)\left(1 - (X - \alpha)^{q-1}\right).$$

14. Let α be a nonzero element of the field $\mathbb{F}_q$ and let k be an integer, with $1 \leq k < q$. We consider the binomial $f(X) = X^k - \alpha$. Demonstrate that the following properties are equivalent :

(i) f divides $X^q - X$,

(ii) $k \mid (q-1)$ and $\alpha^{(q-1)/k} = 1$.

15. The theorem of König-Rados†

Let f be a polynomial with coefficients in the field $\mathbb{F}_q$. We study the number

$$N = \#\{x \in \mathbb{F}_q^* \,;\, f(x) = 0\}.$$

---

† This theorem was shown by J. Koñig and several proofs were given by different authors. One of the first proofs was published by G. Rados in the article Zur Theorie der Congruenzen höheren Grades, *J. reine angew. Math.*, 99, 1886, p.258–260. For more information on this result see the book of Lidl and Niederreiter (*loc. cit.*), Notes of Chapter 6.

1°) Demonstrate that we can only study the case where the degree of f is at most equal to $q-2$. Consequently, we shall suppose, in this exercise that f is given by the formula

$$f = a_0 + a_1 X + \cdots + a_{q-2} X^{q-2}.$$

2°) Let A be the matrix defined by

$$A = (a_{i\,j}) \quad \text{where} \quad a_{i\,j} = a_k, \ k = (j-i) \bmod (q-2).$$

Let $b_1, \ldots, b_{q-1}$ be the elements of $\mathbb{F}_q$ numbered so that we have

$$f(b_j) = 0, \qquad \text{for } j = q-N, \ldots, q-1.$$

Consider the matrix $B = (b_j^{i-1})$. Demonstrate the relation

$$AB = \Big(b_j^{1-i}\, f(b_j)\Big).$$

Demonstrate that the matrix B is invertible, whereas the rank of the matrix AB is equal to $q-1-N$.

3°) Deduce the theorem of König-Rados,

$$N = q - 1 - \operatorname{rank}(A).$$

4°) Numerical application : determine the number of roots in the field $\mathbb{F}_7$ of the polynomial $2X^3 - 3X^2 + X + 3$.

16. The theorem of Chevalley-Warning¶

1°) Let k be a nonnegative integer. Demonstrate that we have

$$\sum_{c \in \mathbb{F}_q} c^k = \begin{cases} 0, & \text{if } k = 0 \text{ or if } k \text{ is not divisible by } (q-1), \\ -1, & \text{otherwise.} \end{cases}$$

(By convention, $0^0 = 1$.)

¶ The original pappers are C. Chevalley, Démonstration d'une hypothèse de M. Artin, *Abh. Math. Sem. Univ. Hamburg*, 11 1936, p. 73-75 — and E. Warning, Bemerkungen zur vorstehenden Arbeit von Herrn Chevalley, *Abh. Math. Sem. Univ. Hamburg*, 11 1936, p. 76-83.

2°) Let $f \in \mathbb{F}_q[X_1, \ldots, X_n]$ be a polynomial of total degree written $\deg(f)$, strictly smaller than $n(q-1)$. Demonstrate the relation

$$\sum_{c_1,\ldots,c_n \in \mathbb{F}_q} f(c_1, \ldots, c_n) = 0.$$

[First, demonstrate the result for a monomial.]

3°) Let F be the polynomial $1 - f^{q-1}$.
Demonstrate that we have

$$F(c_1, \ldots, c_n) = \begin{cases} 1, & \text{if } f(c_1, \ldots, c_n) = 0, \\ -1, & \text{otherwise.} \end{cases}$$

4°) Let $f \in \mathbb{F}_q[X_1, \ldots, X_n]$ be a polynomial of total degree $\deg(f) < q-1$. We put

$$N = \#\{(x_1, \ldots, x_n) \in (\mathbb{F}_q)^n \,;\, f(x_1, \ldots, x_n) = 0\}.$$

Demonstrate the theorem of Warning : $N \equiv 0 \pmod p$, where p is the characteristic of the field $\mathbb{F}_q$. Deduce the theorem of Chevalley : If $f \in \mathbb{F}_q[X_1, \ldots, X_n]$ is a polynomial of total degree $\deg(f) < q-1$ such that $f(0, \ldots, 0) = 0$ then the equation $f(x_1, \ldots, x_n) = 0$ admits a nontrivial solution $(x_1, \ldots, x_n) \in (\mathbb{F}_q)^n$.

5°) Demonstrate that if $f_1, \ldots, f_m \in \mathbb{F}_q[X_1, \ldots, X_n]$ satisfy

$$\deg(f_1) + \cdots + \deg(f_m) < n$$

then, the number

$$N = \#\{(x_1, \ldots, x_n) \in (\mathbb{F}_q)^n \,;\, f_i(x_1, \ldots, x_n) = 0 \text{ for } 0 \le i \le m\}$$

is divisible by p.

[Consider the polynomial $F = (1 - f_1^{q-1}) \cdots (1 - f_m^{q-1})$.]

Deduce a generalization of the theorem of Chevalley.

6°) Let L be the field with q^n elements and let $e_1, \ldots, e_n$ be a basis of L as a vector space over the field $\mathbb{F}_q$. We put

$$N(X_1, \ldots, X_n) = \prod_{k=0}^{n-1} \Big(e_1^{q^k} X_1 + \cdots + e_n^{q^k} X_n \Big).$$

(i) Demonstrate that the polynomial N has coefficients in $\mathbb{F}_q$. [Verify the relation $N^q(X_1, \ldots, X_n) = N(X_1^q, \ldots, X_n^q)$.]

(ii) Demonstrate the implication

$$\Big(x_1, \ldots, x_n \in \mathbb{F}_q \text{ and } N(x_1, \ldots, x_n) = 0 \Big) \implies x_1 = \ldots = x_n = 0.$$

(iii) Compute explicitly examples of such polynomials N for $n = 2$ and $n = 3$ in the case of the field $\mathbb{F}_2$.

17. Let $f \in \mathbb{F}_q[X_1, \ldots, X_n]$ be a nonzero polynomial of degree d and take

$$N = \#\{\, (x_1, \ldots, x_n) \in (\mathbb{F}_q)^n \,;\, f(x_1, \ldots, x_n) = 0 \,\}.$$

1°) Demonstrate the inequality

$$N \leq d\, q^{n-1}.$$

[Argue by induction on n and on d. Distinguish the case where there exists some element c of the field $\mathbb{F}_q$, such that the term $X_1 - c$ divides the polynomial f.]

2°) In the case where f is homogeneous, let N' be the number of nontrivial zeros of f, which means that

$$N' = \#\{\, (x_1, \ldots, x_n) \in (\mathbb{F}_q)^n \,;\, f(x_1, \ldots, x_n) = 0 \text{ and } x_1 \cdots x_n \neq 0 \,\}.$$

Demonstrate the inequality

$$N' \leq d\,(q^{n-1} - 1).$$

18. Let E_d be the set of the $f \in \mathbb{F}_q[X_1, \ldots, X_n]$ of degree at most d and let C be the cardinality of this set.

1°) Demonstrate the formula

$$C = q^{\binom{n+d}{d}}.$$

2°) Let $(c_1, \ldots, c_n) \in (\mathbb{F}_q)^n$ be a fixed point. Demonstrate the formula

$$\#\{\, f \in E_d\,;\; f(c_1, \ldots, c_n) = 0 \,\} = \frac{C}{q^{n-1}}\cdot$$

[Consider the constant term of f.]

3°) For $f \in \mathbb{F}_q[X_1, \ldots, X_n]$, set

$$N(f) = \#\{\, (x_1, \ldots, x_n) \in (\mathbb{F}_q)^n\,;\; f(x_1, \ldots, x_n) = 0 \,\}.$$

Demonstrate the relation

$$\sum_{f \in E_d} N(f) = C\, q^{n-1}.$$

Thus, "on the average", a polynomial of $\mathbb{F}_q[X_1, \ldots, X_n]$ has q^{n-1} zeros in the set $(\mathbb{F}_q)^n$.

4°) Demonstrate the relation

$$\sum_{f \in E_d} \{N(f) - q^{n-1}\}^2 = C\left(q^{n-1} - q^{n-2}\right).$$

[First, demonstrate the formulas

$$\sum_{f \in E_d} N(f)^2 = \sum_{\mathbf{b},\, \mathbf{c} \,\in\, \mathbb{F}_q^n} \;\; \sum_{f \in E_d\,;\; f(\mathbf{b}) = f(\mathbf{c}) = 0} 1\;,$$

and

$$\sum_{f \in E_d} N(f)^2 = \sum_{\mathbf{c} \,\in\, \mathbb{F}_q^n} C/q \;+ \sum_{\mathbf{b},\, \mathbf{c} \,\in\, \mathbb{F}_q^n\,;\; \mathbf{b} \neq \mathbf{c}} C/q^2\;.$$

Thus, "in general", $|N(f) - q^{n-1}|$ is of order of $q^{(n-1)/2}$.

19. Let K be a finite field and let a be an element of K. Demonstrate that there exists x and y in K such that $a = x^2 + y^2$.

[If K is of even characteristic, we can find x such that $a = x^2$. If $K = \mathbb{F}_q$, with q odd, notice that we have $\#\{x^2\,;\, x \in K\} = (q+1)/2 \ldots$]

20. Recall that the resultant of two polynomials

$$f = a\prod_{i=1}^{m}(X - \alpha_i), \quad g = b\prod_{j=1}^{n}(X - \beta_j).$$

can be defined by the formula

$$\operatorname{Res}(f,g) = a^n\, b^m \prod_{i=1}^{m}\prod_{j=1}^{n}(\alpha_i - \beta_j).$$

[Hence $\operatorname{Res}(g,f) = (-1)^{mn}\operatorname{Res}(f,g)$.]

1°) Demonstrate the formula of Swan

$$\operatorname{Res}(X^j - \alpha^j, X^k - \beta^k) = (-1)^j(\beta^m - \alpha^m)^d$$

where $d = \text{g.c.d.}(j,k)$ and $m = \text{l.c.m.}(j,k)$.

[Let R be the value of the sought resultant and let ξ be a primitive jth root of unity. Demonstrate the formulas

$$R = (-1)^j \alpha^{md}\prod_{i=1}^{j}\left\{(\beta/\alpha)^k - \xi^{ki}\right\},$$

and

$$\prod_{i=1}^{j/d}(X - \xi^{ki}) = X^{j/d} - 1.$$

Conclude.]

2°) If f is a monic polynomial of degree $n \geq 1$, we define its discriminant by the formula

$$\operatorname{Discr}(f) = (-1)^{n(n-1)/2} \operatorname{Res}(f, f').$$

Demonstrate the theorem of Swan† :

$$\operatorname{Discr}(X^n + aX^k + b) = \\ (-1)^{n(n-1)/2}\, b^{k-1} \Big\{ n^N b^{N-K} - (-1)^N (n-k)^{N-K}\, k^K\, a^N \Big\}^d,$$

where we have $n > k > 0$, $d = \text{g.c.d.}(n, k)$, $N = n/d$, $K = k/d$.

3°) Let g and h be two monic polynomials, demonstrate the formula

$$\operatorname{Discr}(g \cdot h) = \operatorname{Discr}(g) \cdot \operatorname{Discr}(h) \cdot \big(\operatorname{Res}(g, h)\big)^2.$$

Deduce the general formula

$$\operatorname{Discr}\left(\prod_{i=1}^{k} f_i\right) = \prod_{i=1}^{k} \operatorname{Discr}(f_i) \cdot \left\{ \prod_{1 \leq i < j \leq k} \operatorname{Res}(f_i, f_j) \right\}^2,$$

where $f_1, \ldots, f_k$ are monic polynomials.

4°) In this question, suppose that q is odd. Let $f \in \mathbb{F}_q[X]$ be an irreducible monic polynomial of degree m with one root a in the field with q^m elements. Demonstrate the formula

$$\operatorname{Discr}(f) = \Delta^2 \quad \text{where} \quad \Delta = \prod_{0 \leq i < j \leq m-1} \left(\alpha^{q^i} - \alpha^{q^j}\right).$$

Demonstrate that the discriminant of f is a square in the field $\mathbb{F}_q$ if, and only if, the degree of the polynomial f is even. [Compute Δ^q.]

Using the result of the last question, deduce the theorem of Stickelberger : for q odd, take a monic polynomial $f \in \mathbb{F}_q[X]$, squarefree, of degree equal to m, having k irreducible factors over the field $\mathbb{F}_q$, then, we have

$$k \equiv m \pmod 2 \iff \operatorname{Discr}(f) \text{ is a square in } \mathbb{F}_q.$$

† cf. Berlekamp, *Algebraic Coding Theory*, Th. 6.67.

21. The law of quadratic reciprocity

Let p and q be two distinct odd prime numbers.

1°) If $D = \operatorname{Discr}(X^q - 1)$, demonstrate the formula

$$D = (-1)^{q(q-1)/2} q.$$

2°) Demonstrate that the polynomial $X^q - 1$ has, in the finite field $\mathbb{F}_p$, a number of irreducible factors r given by the formula

$$r = 1 + (q-1)/m,$$

where m is the order of the rational integer p modulo q. Deduce the relation

$$\left(\frac{p}{q}\right) = (-1)^{r+1}.$$

3°) Applying the theorem of Stickelberger (see Exercice 20), demonstrate the formula

$$\left(\frac{D}{p}\right) = (-1)^{q-r}.$$

Demonstrate the relation

$$\left(\frac{D}{p}\right) = \left(\frac{p}{q}\right).$$

4°) Demonstrate the famous "quadratic reciprocity law" :

$$\left(\frac{q}{p}\right) = \left(\frac{p}{q}\right) \cdot (-1)^{\frac{p-1}{2}\frac{q-1}{2}}.$$

(This proof is attributed to Swan‡.)

---

‡ Factorization of polynomials over finite fields, *Pacific J. Math.*, 12, 1962, p. 1099–1106.

22. Let $f \in \mathbb{F}_q[X]$ be a squarefree polynomial and let g be a polynomial of $\mathbb{F}_q[X]$ such that f divides $g^q - g$. Call S the subset of the points of $\mathbb{F}_q$ for which the g.c.d. of the polynomials $f(X)$ and $g(X) - s$ is nontrivial. And set

$$H(X) = \prod_{s \in S} (X - s).$$

Demonstrate that H is the nonzero polynomial of minimal degree such that $f(X)$ divides the composed polynomial $H\big(g(X)\big)$.

(This result is due to H. Zassenhaus*.)

23. Let A be a factorial ring of characteristic p $(p \geq 0)$. If an irreducible polynomial g with coefficients in A divides a polynomial $f \in A[X]$ and we have

$$g^k \,||\, f, \qquad p \text{ does not divide } k, \qquad g' \neq 0,$$

then $g^{k-1} \,||\, f'$ (where the notation $u^k \,||\, v$ means that u^h divides v but u^{h+1} does not divide v).

24. Let $\mathbb{F}_q$ be a finite field of odd characteristic. Prove that the equation

$$w^2 + x^2 + y^2 + z^2 = 0$$

has exactly $(q+1)(q^2-1)+1$ solutions in the set ${\mathbb{F}_q}^4$.

25. Let $f \in \mathbb{F}_q[X]$ be an irreducible polynomial of degree n and let $g \in \mathbb{F}_q[X]$ be a polynomial such that f does not divide any of the polynomials $g^{q^d} - g$, for d dividing n, $1 \leq d < n$. Demonstrate that the congruence

$$h\big(g(X)\big) \equiv X \pmod{F}$$

has a solution h in $\mathbb{F}_q[X]$.

[Consider a root ξ of f in the field L with q^n elements and demonstrate, first, that $g(\xi)$ does not belong to any proper subfield of L. Deduce that there exists one polynomial $h \in \mathbb{F}_q[X]$, such that we have $\xi = h\big(g(\xi)\big)$. Conclude.]

---

* *J. Number theory*, 1, 1969, p. 291-311.

26. Let $f \in \mathbb{F}_q[X]$ be an irreducible polynomial of degree n. Demonstrate that f divides the polynomial $X + X^q + \cdots + X^{q^{n-1}}$.

[Take ξ as in the previous exercise, demonstrate the relation

$$\xi + \xi^q + \cdots + \xi^{q^{n-1}} = 0$$

and conclude.]

27. Let f and $f_1 \in \mathbb{F}_q[X]$ be two irreducible polynomials of degree n. Demonstrate that there exists one polynomial g belonging to $\mathbb{F}_q[X]$ such that

(i) f divides the polynomial $f_1\big(g(X)\big)$,

(ii) f does not divide $f_1'\big(g(X)\big)$,

(iii) f does not divide any of the polynomials g^{q^k} for the integers $k = 1, 2, \ldots, n-1$.

[As in Exercise 25, consider two elements ξ and ξ_1 in the field L that are roots of the polynomials f and f_1, respectively, and note that there exists a polynomial g of $\mathbb{F}_q[X]$ such that $\xi_1 = g(\xi)$ ...]

28. Let p be a prime number. Demonstrate that for any nonzero element a of the field $\mathbb{F}_p$, the polynomial $X^p - X + a$ is irreducible over the field $\mathbb{F}_p$.

[First, demonstrate the congruences $X^{p^n} \equiv na \pmod{f}$ for all nonnegative integers n.]

29. Let n be an integer ≥ 2; and let p be a prime number that does not divide n. Demonstrate that the cyclotomic polynomial Φ_n is irreducible over the finite field $\mathbb{F}_p$ if, and only if, the order of the rational integer p in the group $\mathrm{G}(n)$ is equal to $\varphi(n)$.

30. The theorem of the normal basis

Let K and L be two finite fields, where L includes K, with $\mathrm{Card}(L) = q^n$ and $\mathrm{Card}(K) = q$. Let f be the automorphism of Frobenius of L over K, which means that $f : L \longrightarrow L$ is the function which associates x^q to each element x of L. Let $P(X)$ be the minimal polynomial of f (considered as a K-linear function).

1°) Demonstrate that the polynomial P divides $X^n - 1$.

2°) Demonstrate that the functions Id, f, f^2, …, f^{n-1} are linearly independent over the field K (which is a particular case of a theorem of Dirichlet).

3°) Deduce the equality $P = X^n - 1$.

4°) To any element x of the field L we associate the divisor Q_x of minimal degree of the polynomial P such that we have $Q_x(f)\,(x) = 0$. Demonstrate that if x and y are two elements of L, such that the polynomials Q_x and Q_y are relatively prime, then we have the relation

$$Q_{x+y} = Q_x \cdot Q_y .$$

Demonstrate that there exists an element z of the field L such that we have the relation $Q_z = P$.

5°) Demonstrate that for the element z of the previous question, the family

$$z,\ z^q,\ z^{q^2},\ \ldots,\ z^{q^{n-1}}$$

is a basis of L as vector space over the field K. Such a basis is called a *normal basis.*

6°) If L is represented as K–vector space reported to some normal basis, demonstrate that the computation of the pth power of an element of L is equivalent to a circular permutation on the components of this element.

31. Let A be a ring with a unit element. An element x of A is called an idempotent if $x^2 = x$.

1°) If e is an idempotent, verify that $1 - e$ is also an idempotent.

2°) If e and e' are idempotents such that $e \cdot e' = e' \cdot e = 0$, show that their sum $e + e'$ is also an idempotent.

3°) Let t be an element of A that satisfies $t^q = t$ for some positive integer q.

(i) Suppose that q is odd. If the number 2 is invertible in the ring A, show that the element

$$\frac{1}{2}\left(t^{\frac{q-1}{2}} + t^{q-1}\right)$$

is an idempotent of A.

(ii) Now, suppose that $q-1$ is a multiple of 3 and put $u = t^{\frac{q-1}{3}}$. If the number 3 is invertible in the ring A, show that the element

$$\frac{1}{3}\left(u + u^2 + u^3\right)$$

is an idempotent of A.

Assume that $X^2 + X + 1 = (X-\alpha)(X-\beta)$ for α and β in A. Prove that α and β are distinct. Verify that the element

$$\frac{1}{3}\left(\alpha u + \beta u^2 + u^3\right)$$

is an idempotent of A.

32. Let $f(x_1, \ldots, X_n)$ be a nonzero polynomial over $\mathbb{F}_q$ of total degree d. Demonstrate that the number of zeroes of f in the set ${\mathbb{F}_q}^n$ is at most dq^{n-1}. *

[The result is obvious if $d \leq 1$ or if $n = 1$. Proceed by double induction : suppose $n > 1$, $d > 1$, the result true for polynomials in at most n variables and degree less than d and also the result true for polynomials in less than n variables and of degree at most d. Consider the following cases :

(i) The polynomial f is not divisible by $X_1 - a$ for any a in $\mathbb{F}_q$; then look at the polynomial $f(a, X-2, \ldots, X_n)$.

(ii) We have $f(x_1, \ldots, X_n) = (X_1 - a)\, g(x_1, \ldots, X_n)$ for some a in $\mathbb{F}_q$; then apply the induction hypothesis to g.]

---

* See W. M. Schmidt, *Equations on finite fields*, Lecture Notes in Mathematics, N. 536, Springer, Berlin, 1976, Ch. 4, Sect. 3.

Chapter 7

POLYNOMIALS WITH INTEGER COEFFICIENTS

This chapter is the culmination of this book. Here we describe algorithms of factorization of polynomials with integer coefficients. These algorithms use many of the results demonstrated in the previous chapters.

In this chapter, we consider polynomials with only one variable.

1. Principles of the algorithms of factorization

1. Existence of an algorithm of factorization

Let F be a polynomial with integer coefficients of degree $d \geq 2$ and of height equal to $H = \mathrm{H}(F)$. We always suppose that F is *primitive*, which means that its coefficients are relatively prime.

If the polynomial F is reducible (over $\mathbb{Z}$), then, it admits a nonconstant divisor $G \in \mathbb{Z}[X]$, of degree $d' \leq d/2$. We know how to find upper bounds for the coefficients of such a divisor G. More precisely, we have the inequalities

$$\mathrm{H}(G) \leq \sqrt{d+1}\, 2^{d'} H \leq \sqrt{d+1}\, 2^{d/2} H .$$

Of course, any upper bound of the roots of F leads immediately to an upper bound of the coefficients of the polynomial G. Thus, for example, the inequality of Cauchy gives the upper bound (but this upper bound is not as good as the previous one)

$$\mathrm{H}(G) \leq 2^{d/2} (H+1)^{d/2} .$$

In conclusion, to find a possible factor G as above, we only have to test the polynomials whose degree is at most $d/2$ and whose height is at most $\sqrt{d+1}\,2^{d/2}H$. Since the number of these polynomials is approximately

$$(2\,H)^{d/2}\,e^{d^2/4},$$

this algorithm cannot be used in practice (say, for $d \geq 6$).

2. The algorithm of Kronecker

This algorithm is described in detail in Exercise 3. Now, we only give an example of it. Its principle consists in using the fact, that at any integer point a, the value $G(a)$ of any factor G of the polynomial F must divide the number $F(a)$. Then, we only have to choose a number of points a at most equal to $(d+2)/2$ — since the knowledge of $[(d+2)/2]$ such values determines completely any polynomial G of degree at most $d/2$.

Consider for example the polynomial

$$F(X) = X^5 - 5X^4 + 10X^3 - 10X^2 + 1\,.$$

This polynomial has no roots in $\mathbb{Z}$ since we have

$$F(1) = -3, \quad F(-1) = -25\,.$$

If F is reducible, then, it admits one divisor G of degree two. Moreover — without loss of generality — we can suppose that this divisor has a positive leading coefficient and, hence, we can look for a second-degree factor G,

$$G(X) = X^2 + aX + b\,,$$

where a and b are unknown rational integers.

Since $F(0) = 1$, we must have $b = \pm 1$. Moreover,

$$F(1) = -3 \implies G(1) \in \{\pm 1,\, \pm 3\}\,,$$

that is

$$a = -1 - b + G(1), \qquad \text{where} \;\; G(1) \in \{\pm 1,\, \pm 3\}\,,$$

which gives eight possible candidates for the polynomial G and shows that the coefficient a is odd.

In the same way,

$$F(-1) = -25 \implies G(-1) \in \{\pm 1,\, \pm 5,\, \pm 25\},$$

that is,

$$a = 1 + b - G(-1), \quad \text{where } G(-1) \in \{\pm 1,\, \pm 5,\, \pm 25\}.$$

By subtraction, we find

$$2\,b = G(-1) + G(1) - 2, \quad \text{with } b = \pm 1,$$

which takes away the possibilities $G(1) = \pm 25$, and only leaves the three following cases

$$\text{(i)} \qquad b = -1 \quad \text{and} \quad G(1) = -G(-1) = \pm 1, \quad (a = \pm 1),$$

$$\text{(ii)} \qquad b = 1 \quad \text{and} \quad G(-1) = 1,\ G(1) = 3, \quad (a = 1),$$

$$\text{(iii)} \qquad b = 1 \quad \text{and} \quad G(-1) = 5,\ G(1) = -1, \quad (a = -3).$$

Since $F(2) = -7$, we also have

$$G(2) \in \{\pm 1,\, \pm 7\};$$

hence

$$4 + 2\,a + b = \pm 1 \quad \text{or} \quad \pm 7,$$

which, in case (ii), shows that the only possibility is $a = -1$.

By euclidean division, we note that cases (i) and (ii) never occur. However, the third possibility gives the factorization

$$F(X) = (X^2 - 3\,X + 1)\,(X^3 - 2\,X^2 + 3\,X + 1).$$

Remark

We have not followed the method of Kronecker exactly, because we have taken away some divisors by considering several values of the polynomial F, instead of computing all the possible divisors.

In this example, the number of cases to test was very small because of the small degree of the polynomial G, and mainly because the values of F, that we have used, had very few divisors. This method very quickly becomes extremely expensive when the degree of F increases. This algorithm needs *a priori* a number of tests at least equal to $2^{d/2}$ (this is only because of the number of possible choices for the signs of the values of G at the points of interpolation) — and even much greater than this number in most cases.

3. The principle of modern algorithms

Since the discovery of the algorithm of Berlekamp for the factorization of polynomials over a finite field, all the algorithms of factorization of polynomials with integer coefficients possess the following steps :

step 0 : Reduction to the case of a squarefree polynomial.

step 1 : Let $F \in \mathbb{Z}[X]$ be the studied polynomial. Choose a suitable prime number p and factorize the polynomial f in $\mathbb{F}_p[X]$, where f is the image of F modulo the integer p.

step 2 : Refine the previous factorization of F modulo p into a factorization of F modulo p^n, for a large enough exponent n (say for $p^n \geq B$), using Hensel lemma.

step 3 : Check if this refined factorization gives a true divisor of F in the ring $\mathbb{Z}[X]$.

Our work, is now clear, we must study the following problems :

- choice of the integer p used a modulus,
- choice of the bound B and the construction of Hensel,
- search for the factors in $\mathbb{Z}[X]$.

The two first steps are easy and their study has been prepared in the previous chapters. Nevertheless, it is not the same for the last part where we shall present the recent algorithm called "L^3", which is an abbreviation of the name of its authors A. Lenstra, H. Lenstra and L. Lovász*.

* Factoring polynomials with rational coefficients, *Math. Ann.*, 261, 1982,

2. The choice of the prime modulus

As we have already said, first we factorize f in $\mathbb{F}_p[X]$, where f is equal to F modulo p. In order to apply the algorithm of Berlekamp, we must study the case where f is squarefree.

We test if F is squarefree by computing its g.c.d. with its derivative F'. Using the algorithm of squarefree decomposition studied in the previous chapter, we can assume that F is squarefree (in the ring $\mathbb{Z}[X]$).

The following equivalences hold (in the ring $\mathbb{F}_p[X]$)

$$f \text{ squarefree } \iff \text{g.c.d.}(f, f') = 1 \iff \operatorname{Res}(f, f') \neq 0 .$$

Relatively to the initial polynomial F, the translation is

$$f \text{ squarefree } \iff p \text{ does not divide } \operatorname{Discr}(F) .$$

Since F is supposed to be squarefree, its discriminant Δ is nonzero. It is easy to find an upper bound for $|\Delta|$. By applying the Hadamard inequality, we get, for example

$$|\Delta| \leq ||F||^{d-1} \, ||F'||^{d} \leq d^d \, ||F||^{2d-1} \leq (d+1)^{2d} \, H^{2d-1} ,$$

where we have put $d = \deg(F)$, $H = \mathrm{H}(F)$. (These notations will be kept in the sequel.)

On the other hand, a rational integer $n > 2$ has, at most $1 + \operatorname{Log} n$ distinct prime factors [indeed, a prime divisor different from two is at least equal to three, hence we have the inequality $\omega(n) \leq 1 + (\operatorname{Log} n)/(\operatorname{Log} 3)$].

Thus, the number of prime divisors of the discriminant Δ satisfies

$$\omega(\Delta) = \mathrm{O}(d \operatorname{Log} d + d \operatorname{Log} H) .$$

This proves that there exists a prime number p that does not divide Δ and that satisfies

$$p = \mathrm{O}\big\{(d \operatorname{Log} d + d \operatorname{Log} H) \operatorname{Log}(d \operatorname{Log} d + d \operatorname{Log} H)\big\} .$$

---

p. 515–534.

In fact, a sharper argument, gives the more precise estimate

$$\omega(n) = \mathrm{O}\Big(\frac{\operatorname{Log} n}{\operatorname{Log}\operatorname{Log} n}\Big), \qquad \text{for } n \geq 3,$$

hence, in fact

$$p = \mathrm{O}(d \operatorname{Log} d + d \operatorname{Log} H).$$

See Exercise 5 for a more precise study.

3. Refining the factorization

1. The bound B

Let G be any divisor of the polynomial F in $\mathbb{Z}[X]$. We have seen in Chapter 4 that when the polynomial G is as in

$$G(X) = b_m X^m + b_{m-1} X^{m-1} + \cdots + b_0,$$

its coefficients satisfy the inequalities

$$|b_j| \leq \binom{m}{j} \mathrm{M}(F), \qquad \text{for } j = 0, 1, \ldots, m.$$

Hence, we deduce the bound

$$\mathrm{H}(G) \leq 2^m \mathrm{M}(F), \qquad \text{where } m = \deg(G).$$

From here on we can

- either compute an approximate value of the measure of the polynomial F, for example using the method described in Chapter 4;
- or use directly the inequality $\mathrm{M}(F) \leq ||F||$.

In any case, we always have

$$\mathrm{H}(G) \leq 2^m (d+1)^{1/2} H, \quad m = \deg(G).$$

We compute a factorization of F modulo p^n, the coefficients of the factors being chosen in the interval

$$[-\frac{p^n-1}{2}, +\frac{p^n-1}{2}],$$

and we choose the exponent n large enough so that

$$p^n \geq 2\,\mathrm{H}(G) \quad \text{for any divisor } G \text{ of } F \text{ of degree} \leq d/2.$$

For example, any integer n such that

$$p^n \geq 2^{1+d/2}\,(d+1)^{1/2}\,H\,,$$

satisfies the previous condition. To simplify, we choose such a value for n, which leads to the estimate

$$n = \mathrm{O}\Big(\frac{d + \mathrm{Log}\,H}{\mathrm{Log}\,p}\Big).$$

We put

$$B = 2^{1+d/2}\,(d+1)^{1/2}\,H\,.$$

2. Hensel's lemma

Hensel's lemma has numerous variants. Here, we demonstrate the following result.

THEOREM 7.1 (Hensel). — *Let p be a prime number. Let f, g_0, and h_0 be three polynomials of $\mathbb{Z}[X]$ that satisfy the following properties :*

(i) $\quad g_0$ *is monic and* $\deg(g_0) + \deg(h_0) = \deg(f)$,

(ii) $\quad f(X) - g_0(X)\,h_0(X) \equiv 0 \pmod{p^{1+2k}}$, *where* $p^k \parallel \mathrm{Res}(g_0, h_0)$.

(Recall that the notation $p^k \parallel x$ means that p^k divides x but that p^{k+1} does not divide x.) Then, for any integer $n \geq 0$, there are polynomials g_n and h_n with integer coefficients, such that we have

(1) $\quad g_n$ *is monic and* $\deg(g_n) + \deg(h_n) = \deg(f)$,

(2) $\quad g_n \equiv g_0 \pmod{p^{k+1}}, \quad h_n \equiv h_0 \pmod{p^{k+1}}$,

(3) $\quad f \equiv g_n\,h_n \pmod{p^{n+2k+1}}$.

Remark

Hensel's lemma shows that, if we start from a solution close enough to some relation, then we can find a better solution. This situation is frequent in Numerical Analysis and the method of proof of the Hensel lemma is the similar modulo p of the method of Newton.

We shall give two different demonstrations of Hensel's lemma.

Proof 1 : linear convergence

Starting from g_0 and h_0, we shall build the sequence of the polynomials g_n and h_n. Of course, we argue by induction on the integer n. Note that the case $n = 0$ corresponds exactly to the hypothesis of the theorem. Thus there is nothing to prove in that case.

Suppose $n \geq 1$ and the sequences (g_i) and (h_i) built up to $n-1$. Search the polynomials g_n and h_n given by

$$g_n = g_{n-1} + p^{n+k}\, g_n^*, \quad h_n = h_{n-1} + p^{n+k}\, h_n^*$$

with

$$\deg(g_n^*) \leq \deg(g_0) - 1, \quad \deg(h_n^*) \leq \deg(h_0).$$

Properties (1) and (2) are automatically satisfied. Now, we have to consider condition (3). With the formulas chosen for g_n and h_n, this condition is equivalent to the relation

$$f - g_{n-1}\, h_{n-1} - p^{n+k}\,(g_{n-1}\, h_n^* + h_{n-1}\, g_n^*) \equiv 0 \pmod{p^{n+2k+1}}.$$

If we write

$$f(X) - g_{n-1}(X)\, h_{n-1}(X) = p^{n+2k}\, r_{n-1}(X),$$

we get the equation

$$(*) \qquad g_{n-1}(X)\, h_n^*(X) + h_{n-1}(X)\, g_n^*(X) \equiv p^k\, r_{n-1}(X) \pmod{p^{k+1}}.$$

This equation is equivalent to a linear system where the unknowns are the coefficients of the two polynomials h_n^* and g_n^*.

Let d be the degree of the polynomial f. The number of unknowns of this linear system is the number of coefficients of the polynomials h_n^* and g_n^*; hence, it is

$$\deg(h_n^*) + 1 + \deg(g_n^*) + 1 = \deg(h_0) + 1 + \deg(g_0) = d + 1;$$

and the number of equations of this linear system is $\deg(f) + 1 = d + 1$. Hence, there are as many equations as unknowns.

The determinant associated with this system is the resultant of the polynomials g_{n-1} and h_{n-1}, hence this determinant is divisible by p^k but not by p^{k+1} [after property (ii)]. The formulas of Cramer show that the system

$$g_{n-1}(X)\, h_n^*(X) + h_{n-1}(X)\, g_n^*(X) = p^k\, r_{n-1}(X)$$

admits a solution $(g_n^*,\, h_n^*)$ over the field $\mathbb{Q}$, and also shows that the common denominator (say a) of these two polynomials is not divisible by p. If u is a rational integer such that

$$ua \equiv 1 \pmod{p^{k+1}}$$

then, the previous solution multiplied by ua gives two polynomials with integer coefficients whose reduction modulo p^k satisfies equation $(*)$. This completes the first proof. □

Proof 2 : quadratic convergence

Forget the notations of the first proof. Here we build a sequence of polynomials g_i and h_i that satisfy

$$f - g_i\, h_i \equiv 0 \pmod{p^{2k+2^i}}.$$

We argue by induction on i. For $i = 0$, the polynomials g_0 and h_0 given in the statement of Theorem 7.1 are suitable.

Suppose the polynomials g_i and h_i are built until the order i and search for the polynomials g_{i+1} and h_{i+1} given by formulas as follows

$$g_{i+1} = g_i + p^{2k+2^i}\, g_i^*, \quad h_{i+1} = h_i + p^{2k+2^i}\, h_i^*,$$

again with $\deg(g_i^*) \leq \deg(g_0) - 1$ and $\deg(h_i^*) \leq \deg(h_0)$. Then, we have the relation

$$f - g_{i+1}\, h_{i+1} \equiv f - g_i\, h_i (g_i\, h_i^* + g_i^*\, h_i) \pmod{p^{2k+2^i}},$$

where the hypothesis of induction allows us to write

$$f - g_i\, h_i = p^{2k+2^i}\, r_i, \qquad \text{with } r_i \in \mathbb{Z}[X].$$

We, now, have to solve the equation

$$g_i\, h_i^* + g_i^*\, h_i \equiv p^k\, r_i \pmod{p^{2k+2^i}}.$$

As above, p^k is the greatest power of p that divides the determinant of the linear system corresponding to these relations, and an easy modification of the argument that ends the first proof shows that this equation has, indeed, solutions g_i^* and h_i^*. [The polynomials with rational coefficients considered at the beginning still have, as a common denominator, a rational integer a nondivisible by p; then, we take a rational integer u that satisfies the condition $ua \equiv 1 \pmod{p^{2^i}}$ ...]

This ends the second demonstration. □

The sequence of pairs of polynomials considered in Proof 2 gives approximate solutions of the equation $f - gh = 0$, with a "precision" which is multiplied by approximately two at each step (and which is exactly multiplied by two when k is equal to zero).

COROLLARY 1. — *Let f, g_0, and h_0 be polynomials of $\mathbb{Z}[X]$ that satisfy the three following conditions*

(i) g_0 *and* h_0 *are monic,*

(ii) $f \equiv g_0\, h_0 \pmod{p}$,

(iii) *the polynomials* $(g_0 \bmod p)$ *and* $(h_0 \bmod p)$ *are coprime.*

Then, for every integer $n \geq 0$, there exist polynomials g and h of $\mathbb{Z}[X]$ which satisfy

$$f \equiv g\, h \pmod{p^n},$$

with g monic and

$$g \equiv g_0 \pmod{p}, \qquad h \equiv h_0 \pmod{p}.$$

Corollary 2. — *Let $f \in \mathbb{Z}[X]$ be a polynomial, and let f' be its derivative. Let k and m be two rational integers, such that $0 \leq 2k < m$. Suppose that there exists a rational integer x such that*

$$f(x) \equiv 0 \pmod{p^m} \quad \textit{and} \quad p^k \parallel f'(x).$$

Then, for every rational integer $n \geq m$, there exists a rational integer y, such that

$$f(y) \equiv 0 \pmod{p^n}, \quad p^k \parallel f'(y) \quad \textit{and} \quad y \equiv x \pmod{p^{m-k}}.$$

Proof

By euclidian division, we can write

$$f(X) = f(x) + (X - x)\, f_1(X), \qquad \text{with } f_1 \in \mathbb{Z}[X].$$

Now, we only have to verify that the hypotheses of Theorem 7.1 indeed occur. It is obvious for hypothesis (i) of this theorem.

For hypothesis (ii), first, note that we have the relation (whose proof we leave as an exercise)

$$\operatorname{Res}(g_0, h_0) = \pm f_1(x) \equiv \pm f'(x) \pmod{p^{k+1}},$$

hence $p^k \parallel \operatorname{Res}(g_0, h_0)$, which is the second condition of hypothesis (ii) in Theorem 7.1. From the hypothesis

$$f(x) \equiv 0 \pmod{p^m}, \quad \text{with } m > 2k,$$

of Corollary 2, we obtain the congruence

$$f(X) - (X - x)\, f_1(X) \equiv 0 \pmod{p^{2k+1}},$$

which is the first condition of hypothesis (ii) in Theorem 7.1. We have verified all the hypotheses of Theorem 7.1, which ends the proof. $\square$

4. Berlekamp's method of factorization

Consider a squarefree polynomial $F \in \mathbb{Z}[X]$ of degree d and of height H, which we want to factorize. To simplify, suppose that F monic (for the general case, see Exercise 4). Berlekamp's method contains the following steps :

(1) Choose some prime number p that does not divide the discriminant of the polynomial F.

(2) Factorize the polynomial f in $\mathbb{F}_p[X]$, where f is the image of F modulo p. Let $P_1, \ldots, P_k$ be the irreducible factors of f in $\mathbb{F}_p[X]$.

(3) If $k = 1$, then F is irreducible, and — of course — we stop the procedure.

(4) Otherwise, if $k > 1$, choose any partition of the set $\{1, 2, \ldots, k\}$ in two nonempty parts S and T and put

$$g_0 = \prod_{i \in S} P_i, \qquad h_0 = \prod_{i \in T} P_i,$$

so that we have

$$F \equiv g_0 h_0 \pmod p, \quad \text{with} \quad \operatorname{Res}(g_0, h_0) \not\equiv 0 \pmod p,$$

(consider g_0 and h_0 as polynomials with integer coefficients).

(5) Hensel's lemma can be applied, which allows us to refine the previous congruence into

$$F \equiv gh \pmod{p^n}, \quad p^n \geq B,$$

the bound B is chosen as above, and the polynomials g and h have integer coefficients lying between $-p^{n/2}$ and $p^{n/2}$.

(6) Then, test if g divides F (in the ring $\mathbb{Z}[X]$). At this point, three cases are possible :

- The polynomial g divides F. Hence, we have found a nontrivial factor of the polynomial F. Then, we can apply the present procedure to each of the two factors g and F/g of F that we have just found.

• The polynomial g does not divide F and there exists, at least, one partititon of $\{1,2,\ldots,k\}$ which is not tested. Then, we try such a partition, and go back to step 4.

• All the partitions have been treated : the procedure stops with the conclusion that the polynomial F is irreducible over the integers.

Let us briefly study the cost of this algorithm. The cost of steps 1 to 3 has already been computed, and we know that it is in $\mathrm{O}\left(d^3p\,(\mathrm{Log}\,p)^2\right)$. Moreover, we have seen that we can suppose $p = \mathrm{O}\left(d\ \mathrm{Log}\,(dH)\right)$, hence, the time of computation of the three first steps is at most

$$\mathrm{O}\left(d^4\,(\mathrm{Log}\,dH)^2\right).$$

For any choice of the part S, the following steps have a cost C that satisfy the estimate

$$C = \mathrm{O}\left(d^2\,(d + \mathrm{Log}\,H)^2\right)$$

(prove this estimate as an exercise).

But, we may have to test all the 2^{k-1} possible sets S, then, the cost of the parts 4 to 6 of the algorithm is $2^{k-1}C$. As we have seen in the previous chapter, we "generally" have

$$k \leq 2\,\mathrm{Log}\,d \quad (\text{thus } 2^k < d^{3/2})$$

hence, the algorithm of Berlekamp "generally" has a total cost at most

$$\mathrm{O}\left((\mathrm{Log}\,d)^2\,d^4(\mathrm{Log}\,dH)^2\right).$$

Nevertheless, there are polynomials irreducible over the ring $\mathbb{Z}$ and for which — for any choice of the prime number p — we always have the inequality $k \geq d/2$ (hence $2^{k-1} \geq 2^{d/2-1}$); which shows that the algorithm of Berlekamp is not polynomial with respect to the degree of the polynomial F (see Exercise 5).

The next sections give an algorithm of factorization whose cost is polynomial in function of the degree, the algorithm "L^3".

5. The algorithm L^3

1. Lattices

Let n be a positive rational integer, a subset L of the usual euclidean space $\mathbb{R}^n$ is called a *lattice* if there exists a basis $b_1, \ldots, b_n$ of the space $\mathbb{R}^n$ such that we can write

$$L = \sum_{i=1}^{n} b_i \mathbb{Z} = \Big\{ \sum_{i=1}^{n} r_i b_i \,;\, r_i \in \mathbb{Z},\ 1 \leq i \leq n \Big\}.$$

By definition, the *determinant* of L is

$$\det(L) = |\det(b_1, \ldots, b_n)|.$$

It does not depend on the choice of the basis $b_1, \ldots, b_n$. The integer n is called the *rank* of L.

2. Orthogonalization of Gram-Schmidt

Let $b_1, \ldots, b_n$ be a basis of $\mathbb{R}^n$. From this basis, we calculate an orthogonal basis $b_1^*, \ldots, b_n^*$ of $\mathbb{R}^n$ as follows. The vectors b_i^*, for $1 \leq i \leq n$, and the scalars μ_{ij}, for $1 \leq j < i \leq n$, are recursively defined by the formulas

$$b_i^* = b_i - \sum_{j=1}^{i-1} \mu_{ij} b_j^*, \qquad \mu_{ij} = \frac{(b_i, b_j^*)}{(b_j^*, b_j^*)},$$

where the notation $(\ ,\ )$ designates the usual scalar product in $\mathbb{R}^n$. (As an exercise, verify that these vectors $b_1^*, \ldots, b_n^*$ constitue an orthogonal basis of the space $\mathbb{R}^n$.)

When $b_1, \ldots, b_n$ constitute a basis of a lattice L, after the construction of the b_i^*, the determinant of this lattice satisfies the relation

$$\det(L) = |\det(b_1^*, \ldots, b_n^*)|,$$

and, since $b_1^*, \ldots, b_n^*$ is an orthogonal basis, we also have

$$|\det(b_1^*, \ldots, b_n^*)| = |b_1^*| \cdots |b_n^*|.$$

Hence, in that case,

$$\det(L) = |b_1^*| \cdots |b_n^*|.$$

Notice, also, that

$$|b_i^2| = |b_i^* + \sum_{j=1}^{i-1} \mu_{ij}\, b_j^*|^2 = |b_i^*|^2 + \sum_{j=1}^{i-1} {\mu_{ij}}^2 |b_j^*|^2 \geq |b_i^*|^2, \quad 1 \leq i \leq n.$$

Hence the following result.

THEOREM 7.2 (inequality of Hadamard). — *Let* $b_1,\ b_2,\ \ldots,\ b_n,$ *be* n *nonzero vectors of the euclidean space* $\mathbb{R}^n$, *then*

$$|\det(b_1, \ldots, b_n)| \leq |b_1| \cdots |b_n|,$$

and the equality only occurs when these vectors are two-by-two orthogonal.

3. Reduced basis

We now follow closely the article by A.K. Lenstra, H.W. Lenstra Jr., and L. Lovász†. In the sequel, this article will be denoted [L]. These authors have given the following very important definition :

> a basis $b_1,\ b_2,\ \ldots,\ b_n$ of a lattice of $\mathbb{R}^n$ is said to be *reduced* if it satisfies the following conditions
>
> $$|\mu_{ij}| \leq 1/2, \quad \text{for } 1 \leq j \leq i \leq n$$
>
> and
>
> $$|b_i^* + \mu_{ii-1}\, b_{i-1}^*|^2 \geq \tfrac{3}{4}\, |b_{i-1}^*|^2, \quad \text{for } 1 < i \leq n.$$

PROPOSITION 7.1. — *Let* $b_1,\ b_2,\ \ldots,\ b_n$ *be a reduced basis of a lattice* L *of the space* $\mathbb{R}^n$, *and let* $b_1^*,\ b_2^*,\ \ldots,\ b_n^*$ *be the associated vectors obtained by the procedure of Gram-Schmidt. Then, we have the inequalities*

$$|b_j|^2 \leq 2^{i-1}\, |b_i^*|^2, \quad \textit{for } 1 \leq j \leq i \leq n.$$

---

† Factoring Polynomials with Rational Coefficients, *Math. Ann.*, 261, 1982, p. 515–534.

Proof

According to the definition of a reduced basis, we have

$$|b_i^*|^2 \geq \left(\frac{3}{4} - \mu_{ii-1}^2\right) |b_{i-1}^*|^2 \geq \frac{1}{2}\,|b_{i-1}^*|^2,$$

for $1 < i \leq n$, and we deduce

$$|b_j^*|^2 \leq 2^{i-j}\,|b_i^*|^2, \qquad \text{for } 1 \leq j \leq i \leq n,$$

which implies

$$|b_i|^2 \leq |b_i^*|^2 + \frac{1}{4}\sum_{j=1}^{i-1} 2^{i-j}\,|b_i|^2 \leq 2^{i-1}\,|b_i^*|^2.$$

In fact, we have the inequalities

$$|b_j|^2 \leq 2^{j-1}\,|b_j^*|^2 \leq 2^{i-1}\,|b_i^*|^2,$$

for $1 \leq j \leq i \leq n$. □

Corollary 1. — *With the hypothesis of the above proposition, we have the estimates*

$$\det(L) \leq |b_1| \cdots |b_n| \leq 2^{n(n-1)/4}\det(L),$$

and

$$|b_1| \leq 2^{n(n-1)/4}\det(L)^{1/n}.$$

Proof

The first inequality is Hadamard's. The second inequality is the result of the proposition and of the formula

$$\det(L) = |b_1^*| \cdots |b_n^*|.$$

The third inequality is the result of the proposition and of the second inequality. □

Corollary 2. — *Let L be a lattice and let $b_1,\ b_2,\ \ldots,\ b_n$ be a reduced basis of L. Let $x_1,\ x_2,\ \ldots,\ x_t$ be linearly independent vectors of L. Then, we have the inequalities*

$$|b_j|^2 \le 2^{n-1} \max\{|x_1|^2, |x_2|^2, \ldots, |x_t|^2\}$$

for all the indices $j = 1,\ 2,\ \ldots,\ t$.

Proof

Any nonzero vector x of L can be written as a linear combination of the vectors b_is,

$$x = \sum_{i=1}^{m} r_i\, b_i = \sum_{i=1}^{m} r_i'\, b_i^*,$$

where the r_is are rational integers, and the r_i's are real numbers, with $r_m \ne 0$, and where $m = m(x)$ is some integer that depends on x. Then, we have $r_{m'} = r_m$; hence

$$|x|^2 \ge r_{m'}^2\, |b_m^*|^2 \ge |b_m^*|^2.$$

Put $m(x_j) = m(j)$, for the indices $j = 1,\ 2,\ \ldots,\ t$; then, we have the inequalities

$$|x_j|^2 \ge |b_{m(j)}^*|^2, \quad \text{for } j = 1, 2, \ldots, t.$$

Even if it entails to modify the order of the x_js, we can suppose that we have the inequalities

$$m(1) \le m(2) \le \ldots \le m(t).$$

Since the x_js are linearly independent, it is obvious that we have

$$j \le m(j),$$

for all the indices $j = 1, 2, \ldots, t$. Hence, we have the inequalities

$$|b_j|^2 \le 2^{m(j)-1}\, |b_{m(j)}^*|^2 \le 2^{n-1}\, |b_{m(j)}^*|^2 \le 2^{n-1}\, |x_j|^2,$$

for $j = 1, 2, \ldots, t$, which demonstrates the result. $\square$

4. Algorithm of reduction

This algorithm is detailed in [L], we only give its statement.

```
data : any basis b_1, b_2, ..., b_n of some lattice L ;
output : a reduced basis of L ;
{first step : orthogonalization of Gram-Schmidt}
begin
  for i := 1 to n do
    begin
    |   b*_i := b_i ;
    |   for j := 1 to i - 1 do
    |     begin
    |     |   μ_ij := (b_i, b*_j)/B_j ;
    |     |   b*_i := b*_i - μ_ij bj*
    |     end ;
    |   B_i := (b*_i, b*_i)
    end ;
  k := 2 ;

(i) trans (k - 1) ;
  {the procedure "trans" assures the condition |μ_ij| ≤ 1/2}

  {we now test second condition at rank k}
  if B_k < (3/4 - μ²_{k k-1}) B_{k-1} then
    begin
    |   modif ; {this procedure permutes the vectors b_k and b_{k-1}}
    |   if k > 2 then k := k - 1 ; {go back}
    end ;

  for h := k - 2, k - 3, ..., 1 do trans (h) ;

  if k = n then stop ;
  {if k reaches n, we have get a reduced basis, then stop}
  k := k + 1 ; goto (i) ; {otherwise go forward}
end.
```

algorithm L^3

The procedures "trans" and "modif" are defined below.

```
Procedure trans (h) ;
  if |μ_{kh}| > 1/2 then
    begin
    |   r := [μ_{kh} + 1/2] ; b_k := b_k − r b_h ;
    |   for j := 1 to h − 1 do μ_{kj} := μ_{kj} − r μ_{hj} ;
    |   μ_{kh} := μ_{kh} − r
    end ;
```

Procedure trans

```
Procedure modif ;
  begin
  |   μ := μ_{k k−1} ; B := B_k + m^2 B_{k−1} ;
  |   μ_{k k−1} := μ B_{k−1}/B ;
  |   B_k := B_{k−1} B_k / B ; B_{k−1} := B ;
  |   permute (b_k, b_{k−1}) ;
  |   for j := 1 to k − 2 permute (μ_{kj}, μ_{k−1 j}) ;
  |   for i := k + 1 to n do
  |     begin
  |     |   μ' := μ_{i k−1} − μ μ_{ik} ;
  |     |   μ_{i k−1} := μ_{ik} + μ_{k k−1} μ' ;
  |     |   μ_{ik} := μ'
  |     end ;
  end ;
```

Procedure modif

5. *Cost of the algorithm*

We must, first, show that the algorithm stops. Therefore, Lenstra et al. consider the quantities

$$d_i = \left|\det\left((b_j, b_k)\right)_{1\leq j,k\leq i}\right| = |b_1^*|^2 \cdots |b_i^*|^2, \qquad 0 \leq i \leq n,$$

and $D = d_1 \cdots d_{n-1}$, so that $d_0 = 1$ and $d_n = \det(L)^2$.

During the course of the algorithm, the number D only changes if one of the b_i^*s is modified, which only happens if the condition

$$(*) \qquad |b_k^* + \mu_{k\,k-1}\, b_{k-1}^*|^2 \geq \frac{3}{4}\, |b_{k-1}^*|^2$$

occurs for some integer $k \geq 2$. In this case, the quantity d_{k-1} is multiplied by a factor strictly smaller than $3/4$, whereas the other terms d_i do not change : hence D is reduced by a factor $< 3/4$.

To demonstrate that the algorithm stops, we only have to find some positive lower bound for D, which only depends on the lattice L, and to show, besides, that each passage in the zone (i) of the algorithm leads to the modification of, at least, one of the vectors b_i. Set

$$m(L) = \min\{|x|^2\,;\; x \in L,\; x \neq 0\}.$$

For $i > 0$, the number d_i is equal to the square of the determinant of lattice L_i generated by the vectors $b_1, \ldots, b_i$ in the space $\mathbb{R}\, b_1 + \cdots + \mathbb{R}\, b_i$. We know that any lattice L_i of dimension i contains at least one nonzero vector x_i that satisfies‖

$$|x_i|^2 \leq (4/3)^{(i-1)/2} \left\{\det(L_i)\right\}^{2/i}.$$

We deduce the lower bound

$$d_i \geq (3/4)^{i(i-1)/2} m(L)^i\,.$$

As in the beginning of the procedure, we have

$$d_i \leq |b_1|^2 \cdots |b_i|^2 \leq B^i, \quad \text{where} \quad B = \max\{|b_k|^2\,;\; 1 \leq k \leq n\}.$$

The condition $(*)$ can only happen at most $\mathrm{O}(n^2 \operatorname{Log} B)$ times for each value of the integer k. Then, we easily deduce that the algorithm L^3 does not need more than $\mathrm{O}(n^4 \operatorname{Log} B)$ arithmetical operations, and we can verify that these operations are only on integers of, at most $\mathrm{O}(n \operatorname{Log} B)$ binary digits.

See [L] for the demonstrations.

‖ See, for example, J.W.S. Cassels.— *An introduction to the geometry of numbers*, Springer, Heidelberg, 1971 (Lem. 4, Ch. 1, and Th. 1, Ch. 2).

This work has been further refined especially by the following authors : E. Kaltofen†, A. Schönhage‡, and C. P. Schnorr¶. The progress in these works are somewhat technical. For more details, see these three articles.

6. Factors of polynomials and lattices

In this section, p is a given prime number and k is a fixed positive rational integer. We represent the elements of the set $\mathbb{Z}/p^k\mathbb{Z}$ by integers that belong to the interval $\left]-p^k/2, p^k/2\right]$ and, with this convention, we write $a \mod p^k$ for the natural image of a rational integer a in the set $\mathbb{Z}/p^k\mathbb{Z}$.

For a polynomial given by the formula

$$g = \sum_i a_i X^i \in \mathbb{Z}[X],$$

we set

$$(g \bmod p^k) = \sum_i (a_i \bmod p^k) X^i \in (\mathbb{Z}/p^k\mathbb{Z})[X],$$

Consider a polynomial $f \in \mathbb{Z}[X]$ of positive degree n, and some polynomial $h \in \mathbb{Z}[X]$, of degree d, with $0 < d \leq n$, satisfying the following properties :

(1°) h is monic,

(2°) $(h \bmod p^k)$ divides $(f \bmod p^k)$ in $(\mathbb{Z}/pk\mathbb{Z})[X]$,

(3°) $(h \bmod p)$ is irreducible in $\mathbb{F}_p[X]$,

(4°) $(h \bmod p)^2$ does not divide $(f \bmod p)$ in $\mathbb{F}_p[X]$.

---

† On the Complexity of Finding Short Vectors in Integer Lattices, EUROCAL' 83, London, ed. by van Hulzen, *Lecture Notes in Computer Science*, N. 162, Springer, Heidelberg, 1983.

‡ Factorization of univariate integer polynomials by diophantine approximation and an improved basis reduction algorithm, Proc. of 11th ICALP, Antwerpen, 1984, *Lecture Notes in Computer Science*, N. 172, Springer, Heidelberg, 1984.

¶ A more efficient Algorithm for Lattice Basis Reduction, *J. Algorithms*, 9, 1988, p. 47–62.

PROPOSITION 7.2. — *With the above notations, the polynomial f admits one irreducible factor h_0 in the ring $\mathbb{Z}[X]$ for which $(h_0 \bmod p)$ is a multiple of $(h \bmod p)$ in the ring $\mathbb{F}_p[X]$. Up to the sign, such a polynomial h_0 is unique. Moreover, if g is a divisor of f in $\mathbb{Z}[X]$ then the following properties are equivalent*

(i) $(h \bmod p)$ *divides* $(g \bmod p)$ *in* $\mathbb{F}_p[X]$,

(ii) $(h \bmod p^k)$ *divides* $(g \bmod p^k)$ *in* $(\mathbb{Z}/p^k\mathbb{Z})[X]$,

(iii) h_0 *divides* g *in* $\mathbb{Z}[X]$.

Finally, if these properties hold, then $(h \bmod p^k)$ *divides* $(h_0 \bmod p^k)$ *in the ring* $(\mathbb{Z}/p^k\mathbb{Z})[X]$.

Proof

The existence of h_0, as well as its unicity — up to the sign — do not create any difficulty. Now, let us show the equivalence of the properties (i), (ii) and (iii).

(iii) $\Longrightarrow$(ii) : Clear.

(ii) $\Longrightarrow$ (i) : Obvious.

(i) $\Longrightarrow$ (ii) : Suppose that $f = gq$ in the ring $\mathbb{Z}[X]$ and that the polynomial $(h \bmod p)$ divides $(g \bmod p)$ in the ring $\mathbb{F}_p[X]$. Since the square of $(h \bmod p)$ does not divide $(f \bmod p)$ in $\mathbb{F}_p[X]$, the polynomials $(h \bmod p)$ and $(q \bmod p)$ are relatively prime in $\mathbb{F}_p[X]$; there exist polynomials u, v, and w of $\mathbb{Z}[X]$ such that (in $\mathbb{Z}[X]$) we have the relation

$$u\,h + v\,q = 1 - p\,w.$$

Multiplying this equality by the product $g \cdot \Big(1 + p\,w + \cdots + (p\,w)^{k-1}\Big)$, we get a relation in $\mathbb{Z}[X]$ as

$$u_1\,h + v_1\,f = (1 - p^k w^k)g; \qquad \text{with } u_1,\, v_1 \in \mathbb{Z}[X].$$

Hence property (ii) holds. The last assertion is a particular case of the implication (i) $\Longrightarrow$ (ii). Hence, we have proved the equivalence between properties (i), (ii), and (iii). □

In the sequel of this section we fix a rational integer $m \geq d$, and we call L the set of the polynomials of $\mathbb{Z}[X]$ of degree at most m whose image modulo p^k is divisible by $(h \bmod p^k)$ in the ring $(\mathbb{Z}/p^k\mathbb{Z})[X]$. By identifying a polynomial with the list of its coefficients, we see that L is a lattice of the space $\mathbb{R}^{m+1}$ which admits, as a basis, the set

$$\{p^k X^i\,;\, 0 \leq i < d\} \cup \{hX^j\,;\, 0 \leq j \leq m-d\}.$$

Hence, the determinant of L is $\det(L) = p^{kd}$.

The key to the algorithm of factorization is the following result.

PROPOSITION 7.3. — *Let f and h_0 be polynomials as in the statement of Proposition 7.2. Then, if b is a nonzero element of the lattice L defined above and if b satisfies*

$$|b|^n\,|f|^m < p^{kd},$$

then the polynomial h_0 divides b in the ring $\mathbb{Z}[X]$.

Proof

Let us put $g = \text{g.c.d.}(f, b)$. Using Proposition 7.2, we only have to demonstrate that the polynomial $(h \bmod p)$ divides $(g \bmod p)$ in the ring $\mathbb{F}_p[X]$. We use a *reductio ad absurdum* method of reasoning. Suppose that $(h \bmod p)$ does not divide $(g \bmod p)$. Since $(h \bmod p)$ is irreducible in $\mathbb{F}_p[X]$, there are polynomials u_2, v_2, and $w_2 \in \mathbb{Z}[X]$ such that

$$(*) \qquad u_2\, h + v_2\, g = 1 - p\, w_2\,.$$

Put $e = \deg(g)$, $m' = \deg(b)$; thus $0 \leq e \leq m' \leq m$.

Let L' be the set of linear combinations $u\,f + v\,b + w$ where the polynomials u, v, and w satisfy the following conditions $\deg(u) < m' - e$, $\deg(v) < n - e$ and $\deg(w) < e$. Let us show that L' is a lattice that admits as a basis

$$\{1, X, \ldots, X^{e-1}\} \cup \{X^i f\,;\, 0 \leq i < m' - e\} \cup \{X^j b\,;\, 0 \leq j < n - e\}.$$

Therefore, we only have to prove that a relation as $uf + vb = w$ with $\deg(u) < m' - e$, $\deg(v) < n - e$, $\deg(w) < e$ implies $u = v = w = 0$; which follows immediately from the relation

$$\mathbb{Z}[X] \cdot f + \mathbb{Z}[X] \cdot b = \mathbb{Z}[X] \cdot g,$$

[implied by $g = \text{g.c.d.}(f, b)$], and since g is of degree e.

The inequality of Hadamard implies

$$\det(L') \leq |f|^{m'-e} |b|^{n-e} \leq |f|^m |b|^n < p^{kd}.$$

We shall, now, give a lower bound for the determinant of the lattice L'. Let us show the following implication

$$(**) \qquad z \in L' \text{ and } \deg(z) < e + d \implies \deg(z \bmod p^k) < e.$$

Let z be a polynomial belonging to L' and of degree $< e+d$. By euclidian division,

$$z = qg + r, \qquad \text{with} \quad \deg(r) < e.$$

Multiplying relation $(*)$ by the product

$$q \cdot \left(1 + pw_2 + \cdots + (pw_2)^{k-1}\right),$$

we obtain a formula like

$$u_3 h + v_3 q g \equiv q \pmod{p^k\mathbb{Z}[X]}, \qquad \text{where} \quad u_3, v_3 \in \mathbb{Z}[X].$$

Since b belongs to L', the polynomial $(h \bmod p^k)$ divides $(ug \bmod p^k)$, for some rational integer u; therefore $(h \bmod p^k)$ divises also $(uq \bmod p^k)$. But, on one hand, h is monic of degree d, and, on the other hand, the degree of q is smaller than d. Hence $q = 0$ and we have the implication $(**)$. We can choose a basis $b_0, b_1, \ldots, b_{n+m'-e-1}$ of the lattice L' where each polynomial b_i is of degree i (see, for example, Cassels, *loc. cit.*, Th. 1.A). After the implication $(**)$, the leading coefficient of b_i is divisible by p^k for the indices $i = e, e+1, \ldots, e+d-1$, which implies

$$\det(L') > p^{kd},$$

which is a contradiction. This contradiction ends the demonstration. □

COROLLARY 1. — *We keep the previous notations and suppose the sequence* $b_1, b_2, \ldots, b_{m+1}$ *to be a reduced basis of* L. *We call* M *the measure of the polynomial* f. *Then, we have the two following results :*

(i) $|b_1| < \left(p^{kd}\,|f|^{-m}\right)^{1/n} \implies \deg(h_0) \le m.$

(ii) *If we have the inequality*

$$p^{kd} > 2^{mn/2}\binom{2m}{m}^{n/2}|f|^m M^n,$$

and if the degree of the polynomial h_0 *is at most* m, *then* b_1 *satisfies*

$$|b_1| < \left(p^{kd}\,|f|^{-m}\right)^{1/n}.$$

Proof

Property (i) is an immediate consequence of the proposition since the polynomial b_1 is of degree at most m. To prove (ii), suppose h_0 of degree at most m, then h_0 belongs to L and

$$|h_0|^2 \le \sum_{i=0}^{m}\binom{m}{i}^2 \mathrm{M}(h_0) \le \binom{2m}{m} M.$$

Corollary 2 of Proposition 7.1, applied for $t = 1$ and $x_1 = h_0$, gives

$$|b_1| \le 2^{m/2}|h_0| \le 2^{m/2}\binom{2m}{m}^{1/2} M,$$

hence we obtain property (ii). □

COROLLARY 2. — *Take the notations of the previous corollary and suppose that* p *satisfies the inequality*

(i) $$p^{kd} > 2^{mn/2}\binom{2m}{m}^{1/2}|f|^m M^n,$$

and that there exists one index j, *with* $1 \le j \le m+1$, *for which we have*

(ii) $$|b_j| < \left(p^{kd}\,|f|^{-m}\right)^{1/n},$$

Let t *be the smallest of these indices* j ; *then, we have*

$$\deg(h_0) = m + 1 - t, \quad \text{and} \quad h_0 = \text{g.c.d.}(b_1, b_2, \ldots, b_t),$$

and inequality (ii) holds for all the indices $j \in \{1, 2, \ldots, t\}$.

Proof

Let J be the set of the integers j, where $1 \leq j \leq m+1$, such that inequality (ii) holds. Corollary 1 just above shows us that the polynomial h_0 divides b_j for any index j belonging to J. Hence h_0 divides the gcd of the polynomials b_j, for j running over the set J. Let us write h_1 for this g.c.d. The polynomials b_j, for j belonging to J, are linearly independent, divisible by h_1, and of degree at most m; hence we have the inequality

$$\operatorname{Card}(J) \leq m+1-\deg(h_1).$$

Since the polynomials $X^i h_0$ belong to the lattice L for all the integers $i = 1, 2, \ldots, m-\deg(h_0)$, Corollary 2 of Proposition 7.1 and the upper bound for $|h_0|$, which we have already used, prove that

$$|b_j| \leq 2^{m/2}\binom{2m}{m}^{1/2} M,$$

for $1 \leq j \leq m+1-\deg(h_0)$. Hence all the indices belong to J, which implies the inequality

$$\operatorname{Card}(J) \geq m+1-\deg(h_0).$$

Comparing the lower and upper bounds of $\operatorname{Card}(J)$ found above, we see that we, necessarily, have

$$\deg(h_0) = \deg(h_1) = m+1-t \quad \text{and} \quad J = \{1, 2, \ldots, t\}.$$

Now, we only have to show that h_1 is equal to h_0, up to the sign, therefore, we only have to show that the polynomial h_1 is primitive. Let d_1 be the content of the polynomial b_1. Then the polynomial b_1/d_1 is divisible by h_0, and hence b_1/d_1 belongs to the lattice L. But since the polynomial b_1 is an element of a basis of L, we have $d_1 = 1$. As a conclusion, b_1 is primitive, hence h_1 is also primitive, which ends the demonstration. $\square$

Remark

The proofs of Propositions 7.1 and 7.2 show that the corollary above is true if inequality (i) is replaced by the weaker condition

$$p^{kd} > \beta^n \gamma^n |f|^m,$$

where

$$\beta = \max \{ |b_j|/|b_i^*| \, ; \, 1 \le j \le i \le m+1 \}$$

and

$$\gamma = \max \{ |g| \, ; \, g \in \mathbb{Z}[X], \, g \mid f \}.$$

7. The algorithm of factorization

Consider a squarefree nonconstant polynomial F of $\mathbb{Z}[X]$, primitive and of degree n (suppose also that F is monic, to simplify). Suppose that we have chosen a prime number p, a positive rational integer k and a polynomial $h \in \mathbb{Z}[X]$ was computed to satisfy the properties 1°) to 4°) considered in the previous section.

If h is of degree n, then $(F \bmod p)$ is irreducible, therefore F is also irreducible over $\mathbb{Z}[X]$, and the algorithm stops. This is the reason why we shall suppose also that h is of degree strictly less than n. (For example, we may compute $(h \bmod p)$ with the algorithm of Berlekamp and we refine the computation modulo p^k by applying the method of Hensel.)

Suppose that the coefficients of the polynomial h are reduced modulo p^k (that is, are lying between $-p^k/2$ and $p^k/2$), so that

$$|h|^2 \le 1 + d\,p^k/2, \qquad \text{where } \ d = \deg(h).$$

Consider a given rational integer m, where $m \geq d$, and suppose that p and k satisfy the inequality

$$\text{(i)} \qquad p^{kd} > 2^{mn/2} \binom{2m}{m}^{n/2} |F|^m M^n,$$

(see Corollary 2 of Proposition 7.3). Let L be the lattice considered in the previous section, of basis

$$\{p^k X^i \,;\, 0 \leq i < d\} \cup \{hX^j \,;\, 0 \leq j < m-d\}.$$

First,compute a reduced basis $b_1, b_2, \ldots, b_{m+1}$ of L thanks to the algorithm of Lenstra et al.

After the second corollary of Proposition 7.2, we have

- if $|b_1| \geq \left(p^{kd} |F|^{-m}\right)^{1/n}$ then $\deg(h_0) > m$,
- if $|b_1| < \left(p^{kd} |F|^{-m}\right)^{1/n}$, then $\deg(h_0) \leq m$, as well as

$$h_0 = \text{g.c.d.}\,(b_1, b_2, \ldots, b_t),$$

where the integer t is defined as in the second corollary of Proposition 7.2, which ends the description of the algorithm.

This algorithm gives an irreducible factor h of the polynomial F in $\mathbb{Z}[X]$. For a complete factorization of F in $\mathbb{Z}[X]$, we can, for example, proceed as follows.

- Completely factorize the polynomial $(F \bmod p)$, the integer p being a prime number that does not divide the discriminant of F.

- Choose a polynomial h_1 of $\mathbb{Z}[X]$ such that $(h \bmod p)$ is an irreducible factor of $(F \bmod p)$.

- Determine, as we have just said, a polynomial $h_{1,0}$ of $\mathbb{Z}[X]$ that is an irreducible factor of F in $\mathbb{Z}[X]$.

- If the degree of the polynomial $h_{1,0}$ is strictly smaller than the one of F, choose a polynomial of $\mathbb{Z}[X]$ whose image modulo p is an irreducible factor of the polynomial $(F/h_{1,0} \bmod p)$ in the ring $\mathbb{F}_p[X]$, and apply the same algorithm again with $F/h_{1,0}$ as new polynomial, and so on, until the considered polynomial considered is irreducible in $\mathbb{Z}[X]$.

Considering the cost of the algorithm of Lenstra et al. and studying the auxiliary necessary computations, it is possible to demonstrate the following result. (See [L] for a proof.)

THEOREM 7.2. — *The algorithm described above factorizes any primitive polynomial with integer coefficients into a product of polynomials irreducible in* $\mathbb{Z}[X]$. *This algorithm needs a number of computations at most*

$$\mathrm{O}(n^6 + n^5 \operatorname{Log} |F|),$$

these computations are realized with rational integers of binary length bounded above by

$$\mathrm{O}(n^3 + n^2 \operatorname{Log} |F|).$$

Exercises

1. Let $F \in \mathbb{Z}[X]$ be a polynomial such that $F(0)$ and $F(1)$ are odd. Demonstrate that this polynomial has no root among the rational integers. What about roots in $\mathbb{Q}$?

2. Let p be a prime number and let f be a polynomial with integer coefficients. Demonstrate that the two following properties are equivalent :

(i) $\forall a \in \mathbb{Z},\ f(a) \equiv 0 \pmod{p}$,

(ii) $\exists g, h \in \mathbb{Z}[X],\ f(X) = (X^p - X)\, g(X) + p\, h(X)$.

3. Consider the following procedure, due to Kronecker, to factorize polynomials with integer coefficients.

```
data : f ∈ Z[X], deg(f) = d ≥ 1.
for s = 1, 2, ..., [d/2] do
(i)   choose s + 1 distinct integers x_0, x_1, ..., x_s ;
(ii)  if there exists x_i such that f(x_i) = 0
          then return g = X − x_i and stop
          else determine the set E of (s + 1)-uples of integers
               (d_0, d_1, ..., d_s) ∈ Z^{s+1} such that each d_i runs
               over the set of the divisors of f(x_i) ;
(iii) for all d = (d_0, d_1, ..., d_s) ∈ E compute the polynomial
      g_d ∈ Z[X] such that g_d(x_i) = d_i for i = 0, 1, ..., s ;
      if g_d divides f (in Q[X]) then return g_d and stop.
```

Kronecker's method of factorization

Demonstrate that if this procedure stops without having returned some polynomial g then f is irreducible over $\mathbb{Q}[X]$. Using this method, demonstrate that the polynomial

$$f(X) = 2X^5 + 8X^4 - 7X^3 - 35X^2 + 12X - 1$$

admits the decomposition

$$f(X) = (2X^2 + 6X - 1)\,(X^3 + X^2 - 6X + 1).$$

4. Demonstrate that, for all n, the cyclotomic polynomial Φ_n is irreducible over the ring $\mathbb{Q}[X]$.

[First, demonstrate that it is sufficient to prove the irreducibility of Φ_n over $\mathbb{Z}$. Then consider the minimal polynomial f over $\mathbb{Z}$ of the complex number $\zeta = e^{2i\pi/n}$. Then f divides Φ_n. Let p be a prime number that does not divide the integer n and let g be the minimal polynomial over $\mathbb{Z}$ of ζ^p; then f divides $g(X^p)$ and — when g is different from f — the polynomial Φ_n has a double root modulo p. Show that this is not the case and conclude.]

5. Choice of the prime modulus p

Like in Section 2 above, we are looking for an upper bound of the smallest prime number that does not divide some given rational integer N (indeed, we are interested in the case where N is the absolute value of the discriminant of a certain polynomial with integer coefficients).

1°) Let n be a rational integer ≥ 2. Consider the binomial coefficient

$$C_n = \binom{2n}{n}.$$

Demonstrate the following implications :

(i) $n < p \leq 2n \Longrightarrow p \mid C_n$,

(ii) $3n/2 < p \leq n \Longrightarrow p \nmid C_n$,

(iii) $p^h \mid Cn \Longrightarrow p^h \leq 2n$.

Deduce the formula

$$(*) \qquad 0 \leq \operatorname{Log} C_n - \sum_{n<p\leq 2n} \operatorname{Log} p \leq \sum_{\sqrt{2n}<p\leq 3n/2} \operatorname{Log} p + \rho,$$

where ρ satisfies $0 \leq \rho \leq \sum_{p\leq\sqrt{2n}} \operatorname{Log} n$.

2°) Demonstrate the relation

$$\operatorname{Log} C_n = 2n \operatorname{Log} 2 + \mathrm{O}(\operatorname{Log} n).$$

3°) For any positive real number x, we put

$$\theta(x) = \sum_{p \le x} \operatorname{Log} p.$$

Using the first inequality of $(*)$, demonstrate the estimate

$$\theta(x) \le 2x \operatorname{Log} 2 + \mathrm{O}(\operatorname{Log} x).$$

6. Suppose that F is a polynomial with integer coefficients and of leading coefficient equal to a, where $a > 1$. Demonstrate that to factorize the polynomial F one may consider the polynomial $G = aF$ and just look for the divisors of G, the leading coefficient of which is equal to a.

7. Study the factorization of the polynomial $X^4 + 1$ in all the fields $\mathbb{F}_p$, and demonstrate that its factors irreducible over such a field are of degree at most equal to two. Nevertheless, show that this polynomial is irreducible over $\mathbb{Z}[X]$.

8. Let n be a positive rational integer and let $p_1, \ldots, p_n$ be distinct prime numbers.

1°) Demonstrate that there exists no relation like

$$\sum_{\sigma} a_\sigma \sqrt{\pi_\sigma} = \sqrt{p_n},$$

where σ runs over the subsets of $\{1, 2, \ldots, n-1\}$, and where

$$\pi_\sigma = \prod_{k \in \sigma} p_k.$$

[Argue by induction on n.]

2°) Demonstrate that the polynomial $f(X)$ defined by

$$f(X) = \prod_{\varepsilon_i \in \{0,1\}} \Big(\sum_{i=1}^{n} \varepsilon_i \sqrt{p_i}\Big)$$

has integer coefficients. Using the previous question, demonstrate that f is irreducible over $\mathbb{Z}$.

2°) Demonstrate that for all prime numbers p, the polynomial $f \bmod p$ splits over the field $\mathbb{F}_p[X]$ into a product of irreducible factors of degree ≤ 2. †

9. Eisenstein criterion of irreducibility

Let

$$f(X) = X^n + a_{n-1}X^{n-1} + \cdots + a_1 X + a_0$$

be a polynomial with integer coefficients such that there exists a prime number p dividing each of the integers a_i but such that p^2 does not divide a_0. Demonstrate that f is irreducible over the ring $\mathbb{Q}[X]$. As an application, prove that the cyclotomic polynomial

$$\Phi_p(X) = X^{p-1} + \cdots + X + 1$$

is irreducible over $\mathbb{Q}[X]$.

[Consider the polynomial $\Phi_p(X+1)$.]

10. Hasse principle

1°) Show that there exists a linear polynomial with integer coefficients which has no root among the rational integers but has at least one root in every finite field. [It is possible to take, for example, the linear polynomial $2X + 4$.]

---

† This example is due to H.P.F. Swinnerton-Dyer. See the article of E. Kaltofen, Factorization of Polynomials in *Computer Algebra*, ed. B. Buchberger, G. E. Collins, R. Loos, Springer, 1982, p. 95–113.

2°) Let a and b be two integers. Demonstrate that if the congruence

$$ax + b \equiv 0 \pmod{m}$$

is solvable for all moduli m then the equation $ax + b = 0$ admits an integer solution. [Consider the modulus $m = a$.]

3°) Demonstrate that the congruence

$$6x^2 + 5x + 1 \equiv 0 \pmod{m}$$

is solvable for all moduli m, but has no solution in $\mathbb{Z}$.

[For the first assertion, factorize the polynomial $6x^2 + 5x + 1$ over $\mathbb{Z}$, then apply the Chinese remainder theorem.]

11. We have seen, in Chapter 6, that a random polynomial of degree n over a finite field is reducible with a probability close to $1 - 1/n$. Using the Chinese remainder theorem, show that this implies that a random polynomial of $\mathbb{Z}[X]$ is "almost always" irreducible.

12. Let E be the $\mathbb{Z}$-module of polynomials $P \in \mathbb{Q}[X]$ such that $P(n)$ is an integer for any rational integer n. Demonstrate that the set of polynomials $(P_n)_{n \geq 0}$ defined by

$$P_0(X) = 1, \; P_n(X) = \frac{1}{n!}\, X(X+1)\cdots(X+n-1), \quad n \geq 1$$

constitutes a basis of E over $\mathbb{Z}$.

13. Factorization over $\mathbb{Z}[X]$ and over $\mathbb{Z}$

1°) Let $F(X)$ be a polynomial with integer coefficients and of height at most equal to A. Suppose that a is an integer such that $|a| > 2A$. Demonstrate that the knowledge of the value $F(a)$ completely determines the polynomial F.

2°) Let F and G be two polynomials of $\mathbb{Z}[X]$, and suppose that the height of any common divisor of these two polynomials is at most A. Let a be a rational integer, $|a| > 2A$. Set $f = F(a)$ and $g = G(a)$ and assume that h is the g.c.d. of the integers f and g. Let H be the unique polynomial of height $\leq A$ defined by $H(a) = h$. Prove that H is the g.c.d. of the polynomials F and G if, and only if, it divides both F and G.

APPENDIX

DETERMINANTS

Determinants were used at several places and we find it convenient to recall here their main properties.

In the sequel R is a commutative ring. We denote $M_{m,n}(R) = M_{m,n}$, the set of matrices of size $m \times n$ with coefficients in R. The set of square matrices of size $n \times n$ is denoted $M_n(R) = M_n$.

1. Definition

If $A = (A_{ij})$ is a square matrix of size $n \times n$, the determinant of A is defined recursively by the formulas :

$$\det(A) = |A| = \begin{cases} a_{11}, & \text{if } n = 1, \\ \sum_{i=1}^{n} (-1)^{i+1} a_{i1} \det(A_{i1}), & \text{otherwise,} \end{cases} \tag{1}$$

where A_{ij} denotes the matrix obtained from A by deleting the i-th row and the jth column.

This formula is called the *Laplace expansion* relative to the first column.

Example

$$\begin{vmatrix} X_{11} & X_{12} \\ X_{21} & X_{22} \end{vmatrix} = X_{11} X_{22} - X_{12} X_{21}.$$

2. Laplace expansions

The formulas of this section generalize formula (1).

The Laplace expansion relative to the jth column is the formula

$$\det(A) = \sum_{i=1}^{n} (-1)^{i+j} a_{ij} \det(A_{ij}), \tag{2}$$

and the Laplace expansion relative to the ith column is

$$\det(A) = \sum_{j=1}^{n} (-1)^{i+j} a_{ij} \det(A_{ij}). \tag{2a}$$

Below we shall see that all these formulas give the same value for the determinant.

3. Expansion of the determinant

When σ is a permutation of some finite interval of the integers, we define the *signature*, denoted $\varepsilon(\sigma)$ or sometimes $\operatorname{sgn}(\sigma)$, of this permutation by the formulas

$$\varepsilon(Id) = 1, \quad \varepsilon(\tau) = -1, \quad \text{and} \quad \varepsilon(\sigma\sigma') = \varepsilon(\sigma)\varepsilon(\sigma'),$$

where τ is a transposition and σ' is any permutation.

The determinant satisfies the relation

$$\det(A) = \sum_{\sigma} \varepsilon(\sigma) \prod_{i=1}^{n} a_{i\,\sigma(i)}. \tag{3}$$

Proof

This formula is a consequence of Eq. (1) [use induction on n]. Then, it is easy to verify that Eq. (3) implies formulas (2) and (2a). Thus, as asserted before, these two formulas give the same value as formula (1). $\square$

If $V_1, V_2, \ldots, V_n$ are column vectors of R^n, then, by definition, the determinant of these n vectors $\det(V_1, V_2, \ldots, V_n)$ is the determinant of the matrix of size $n \times n$ whose jth column is equal to V_j, for $1 \leq j \leq n$.

4. Case of two equal columns

If two vectors of the list $V_1, V_2, \ldots, V_n$ are equal then the determinant of these n vectors is zero, that is

$$\det(V_1, \ldots, V_i, \ldots, V_i, \ldots, V_n) = 0.$$

Proof

Use induction on n. The example above shows that this result is true for $n = 2$. Suppose that $n > 2$ and that the result is true for $n - 1$ vectors. Then, suppose that $V_i = V_j$, $i \neq j$, and use the Laplace expansion of the determinant relative to the kth column, for some $k \neq i, j$. The conclusion follows by the induction hypothesis. ☐

5. *Multilinearity*

The function $f : (V_1, V_2, \ldots, V_n) \longmapsto \det(V_1, V_2, \ldots, V_n)$ *is multilinear, that is*

$$\det(V_1, \ldots, \lambda V_i' + \mu V_i'', \ldots, V_n) \\ = \lambda \det(V_1, \ldots, V_i', \ldots, V_n) + \mu \det(V_1, \ldots, V_i'', \ldots, V_n),$$

for each i, *with* $1 \leq i \leq n$, *and any* λ, μ *in* R.

Proof

Use Eq. (3). ☐

6. *Alternate function*

For indices $1 \leq i < j \leq n$, *we have*

$$\det(V_1, \ldots, V_{i-1}, V_j, V_{i+1}, \ldots, V_{j-1}, V_i, V_{j+1}, \ldots, V_n) \\ = -\det(V_1, \ldots, V_i, \ldots, V_j, \ldots, V_n),$$

In other words, the function f defined in Section 5 is alternate.

Proof

Compute the expression $\det(V_1, \ldots, V_{i-1}, V_i + V_j, V_{i+1}, \ldots, V_{j-1}, V_i + V_j, V_{j+1}, \ldots, V_n)$. Using Section 4 and the fact that the function f is additive, we get

$$\det(V_1, \ldots, V_i + V_j, \ldots, V_i + V_j, \ldots, V_n) = 0 \\ = \det(V_1, \ldots, V_i \ldots, V_j, \ldots, V_n) + \det(V_1, \ldots, V_i, \ldots, V_i, \ldots, V_n) \\ + \det(V_1, \ldots, V_j, \ldots, V_i, \ldots, V_n) + \det(V_1, \ldots, V_j, \ldots, V_j, \ldots, V_n) \\ = \det(V_1, \ldots, V_i \ldots, V_j, \ldots, V_n) + \det(V_1, \ldots, V_j, \ldots, V_i, \ldots, V_n);$$

hence we have the result. ☐

7. Case of dependent vectors

If any of the vectors $V_1, V_2, \ldots, V_n$ is equal to a linear combination of the other ones then the determinant of these vectors is zero.

Proof

Using Section 6, we may suppose, without loss of generality, that V_n is a linear combination of the vectors $V_1, V_2, \ldots, V_{n-1}$. Then, using multilinearity, we see that the determinant is a linear combination of determinants that all contain some vector which is repeated. Hence, using Section 4, we see that these determinants are all zero. □

8. Transposition

The determinant is invariant under transposition, that is

$$\det({}^tA) = \det(A).$$

Proof

The proof is an easy consequence of Section 3. □

9. Determinant of a triangular matrix

Without loss of generality, we may suppose that A is an upper triangular matrix (using Section 8, if necessary, transpose A).

If A is an upper triangular matrix,

$$A = T = \begin{pmatrix} \lambda_1 & & \\ 0 & \ddots & * \\ & & \lambda_n \end{pmatrix},$$

then

$$\begin{vmatrix} \lambda_1 & & \\ 0 & \ddots & * \\ & & \lambda_n \end{vmatrix} = \lambda_1 \cdots \lambda_n .$$

Proof

Use rule (1) and induction on n. □

Remark

Formula (1) is convenient to define the determinant but does not lead to an efficient way to compute it. For a square matrix of size n, the evaluation of this formula is very expensive : it needs more than $n\,!$ operations (except, maybe, for "sparse matrices"). Whereas a triangularization of A leads to a method to compute the determinant of A in $\mathrm{O}(n^3)$ operations (this is Gauss' method).

10. Case of independent vectors

If R is an integral domain and if the vectors V_1, V_2, ..., V_n are linearly independent, then the determinant of these vectors is not zero.

Proof

First, notice that we may realize all the computations in the quotient field of R. In other words, in this section, we suppose that R is a field.

We prove the result using induction on n. If $n = 1$, the result is trivial. Now suppose $n > 1$. Let x_i be the first component of V_i for $1 \leq i \leq n$. Without loss of generality, we may suppose that x_1 is non zero. Using properties shown in Sections 5 and 9, we see that the value of the determinant does not change if we replace V_i by $V_i - (x_i/x_1)\,V_1$ for $2 \leq i \leq n$. Then, using Laplace expansion relative to the first row, we can use the induction hypothesis to conclude. □

11. Cauchy-Binet formula

Before we can state the Cauchy-Binet formula, we have to introduce some notations.

If A is a matrix of size $m \times n$ and if $I = \{i_1, \ldots, i_r\}$ (respectively if $J = \{j_1, \ldots, j_s\}$) is a subset of $\{1, 2, \ldots, m\}$ (respectively a subset of $\{1, 2, \ldots, n\}$) then the notation $A(I, J)$ represents the matrix obtained from A by eliminating all rows except rows i_1, ..., i_r and by eliminating all columns except columns j_1, ..., j_s.

(We suppose $i_1 < \ldots < i_r$ and $j_1 < \ldots < j_s$.)

Then the Cauchy-Binet formula is the following result.

If $A \in M_{m,k}$, $B \in M_{k,n}$, and $C = AB$, then for $r \leq \min\{m,k,n\}$, and for any two sets I and J both of cardinality r with $I \subset \{1,2,\ldots,m\}$ and $J \subset \{1,2,\ldots,n\}$, we have the formula

$$\det(C) = \sum_{K \in \mathcal{P}_r\{1,2,\ldots,k\}} \det A(I,K) \det B(K,J),$$

where $\mathcal{P}_r\{1,2,\ldots,k\}$ is the set of subsets of $\{1,2,\ldots,k\}$ which contain exactly k elements.

Proof

By definition of the matrix product,

$$C = (c_{ij}), \quad \text{where} \quad c_{ij} = \Big(\sum_{\nu=1}^{k} a_{i\nu} b_{\nu j}\Big).$$

Thus,

$$\det C(I,J) = \sum_{\sigma} \varepsilon(\sigma) \Big(\sum_{\nu_1=1}^{k} a_{i_1 \nu_1} b_{\nu_1 \sigma(j_1)}\Big) \cdots \Big(\sum_{\nu_k=1}^{k} a_{i_1 \nu_k} b_{\nu_k \sigma(j_1)}\Big),$$

where σ runs over the permutations of the set J.

Upon expanding, we get a sum of n^k terms as follows

$$\begin{aligned}\det M(I,J) &= \sum_{\nu}\sum_{\sigma} \varepsilon(\sigma) \Big(a_{i_1 \nu_1} b_{\nu_1 \sigma(j_1)} \cdots a_{i_k \nu_k} b_{\nu_k \sigma(j_r)}\Big) \\ &= \sum_{\nu} a_{i_1 \nu_1} \cdots a_{i_r \nu_r} \sum_{\sigma} \varepsilon(\sigma) \Big(b_{\nu_1 \sigma(j_1)} \cdots b_{\nu_r \sigma(j_r)}\Big),\end{aligned}$$

where $\nu = (\nu_1, \ldots, \nu_r)$ with $1 \leq \nu_1, \ldots, \nu_r \leq n$.

In the last parenthesis, we recognize the expansion of some determinants, up to the signs. From Section 4, note that all these determinants are zero when two indices ν_i are equal. Thus, we suppose that these indices in the summation are all distinct. Now, taking care of the signs, we get

$$\det M(I,J) = \sum_{K}\sum_{\rho} a_{i_1 \rho(i_1)} \cdots a_{i_r \rho(i_r)} \varepsilon(\rho) \det B(I,J),$$

where K runs over $\mathcal{P}_r\{1,2,\ldots,k\}$ and where ρ runs over the permutations of the set K. Using once more formula (3), we get the result. $\square$

Remark

The special case $m = n$ of the Cauchy-Binet formula was proved by Laplace in 1772. The formula of Cauchy-Binet was proved independently by Cauchy and Binet in 1812.

12. Determinant of a product of matrices

When A and B are two square matrices then — as a special case of the Cauchy-Binet formula — we have the relation

$$\det(A\,B) = \det(A)\,\det(B).$$

In particular, when A is invertible, we have

$$\det(A\,A^{-1}) = \det(A)\,\det(A^{-1}) = 1,$$

which shows that $\det(A)$ *is a unit of the ring R and that*

$$\det(A^{-1}) = \det(A)^{-1}.$$

13. Cramer's rule

Consider a linear system $AX = B$ where $A \in M_n(R)$ is an invertible matrix and X and B are column vectors. Let V_i, $1 \le i \le n$, be the columns of the matrix A. Then, for any i, $1 \le i \le n$, we have

$$\begin{aligned}\det(V_1, \ldots, V_{i-1}, \sum_{k=1}^{n} x_k\,V_k, V_{i+1}, \ldots, V_n) &= x_i\,\det(A)\\ &= \det(V_1, \ldots, V_{i-1}, B, V_{i+1}, \ldots, V_n).\end{aligned}$$

We have just seen that $\det(A)$ is an invertible element of the ring R. Thus, we get Cramer's formulas (obtained in 1750).

The components $x_1, \ldots, x_n$ of the vector X are given by the formulas

$$x_i = \frac{\det(V_1, \ldots, V_{i-1}, B, V_{i+1}, \ldots, V_n)}{\det(A)}, \quad \textit{for} \quad 1 \le i \le n.$$

14. Compound matrices

If $A \in M_{m,n}$ and if k is any positive integer, the *k–compound* of A is the matrix whose entries are all the minors of size k extracted from A. This matrix is denoted $C_k(A)$. [This is a matrix of size $\binom{m}{k} \times \binom{n}{k}$, and we choose the lexicographical order for the indices, which are sequences of integers.]

Using this definition, the formula of Cauchy-Binet is written

$$(*) \qquad C_r(AB) = C_r(A)\, C_r(B).$$

Thus, when A is invertible, we have the formula

$$C_r(A^{-1}) = C_r(A)^{-1}.$$

We also have the obvious relations

$$C_r(\lambda A) = \lambda^r\, C_r(A), \quad C_r({}^tA) = {}^tC_r(A), \quad C_r(I) = I \;(\in M_{\binom{m}{k}}).$$

One of the main interests in the compound matrix is the following fact.

Suppose that A is a square matrix of size $n \times n$, over an integral domain. Let $\lambda_1, \lambda_2, \ldots, \lambda_n$ be the family of the eigenvalues of A (in some field extension F containing R). Then the eigenvalues of the matrix $C_k(A)$ in F are the products $\lambda_{i_1} \cdots \lambda_{i_k}$ for all $(i_1, i_2, \ldots, i_k)$ with $1 \le i_1 < \ldots < i_k \le n$.

Proof

We use the method of algebraic identities (see McDonald, p. 24). Thus we suppose that R is the field of complex numbers. Then, by an argument of continuity, we may suppose that matrix A is diagonizable. Using relation $(*)$, we may even suppose that A is a diagonal matrix, say $A = D(\lambda_1, \ldots, \lambda_n)$ [with obvious notations!]. Now, it is easy to verify that the k–compound matrix of A is equal to

$$C_k(A) = D(\lambda_1 \cdots \lambda_k, \ldots, \lambda_{n-k+1} \cdots \lambda_n).$$

Hence the result. $\square$

Exercise

Let P be a monic polynomial with complex coefficients,

$$P = X^d + a_{d-1}X^{d-1} + \cdots + a_0 = (X - \alpha_1)\cdots(X - \alpha_d).$$

Considering the different compounds of the companion matrix of P prove the following inequalities†

$$|\alpha_1\alpha_2\cdots\alpha_n| \leq nH + 1 \quad \text{for all} \quad n = 1, 2, \ldots, d,$$

where H is the height of P, that is,

$$H = \max\left\{1, |a_{d-1}|, \ldots, |a_0|\right\}.$$

15. Adjoint matrix

Let $A = (a_{ij})$ be an $n \times n$ matrix. As in Section 1, we denote A_{ij} the square matrix obtained from A by deleting the ith row and the jth column. If

$$B = (b_{ij}), \quad \text{where} \quad b_{ij} = (-1)^{i+j} \det(A_{ij})$$

the matrix ${}^tB$ is called the *adjoint* of A, and we denote it $\operatorname{adj}(A)$.

Using Sections 1 and 4, we get the important formula.

We have

$$A \cdot \operatorname{adj}(A) = \operatorname{adj}(A) \cdot A = \det(A) \cdot I.$$

In particular, when A is invertible, then $\det(A)$ *is a unit in the ring R and*

$$A^{-1} = (\det A)^{-1} \operatorname{adj}(A).$$

† This result is due to W. Specht. — Zur Theorie der algebraischen Gleichungen, *Jahresber. Deutsch. Math.*, 5, 1938, p. 142–145.

Notice also the formula

$$\operatorname{adj}(A) = E\,C_{n-1}(A)\,{}^tE,$$

where

$$E = \begin{pmatrix} 1 & & & & \\ & -1 & & 0 & \\ & & 1 & & \\ & 0 & & \ddots & \\ & & & & (-1)^n \end{pmatrix}.$$

16. Schur complement

Let A be a square of size $n \times n$ and let J be a nontrivial subset of $\{1, 2, \ldots, n\}$, such that the matrix $A(J)$ is invertible. Then, Schur proved the following formula :

$$\det(A) = \det A(J)\,\det\Big(A(J') - A(J', J)\,A(J)^{-1}\,A(J, J')\Big).$$

The term $A(J', J)\,A(J)^{-1}\,A(J, J')$ is called the Schur complement of $A(J)$ in A.

Proof

Without loss of generality, we may suppose that $J = \{1, 2, \ldots, k\}$. Put $A_0 = A(J)$ and $A_1 = A(J')$. Then A may be written

$$A = \begin{pmatrix} A_0 & B \\ C & A_1 \end{pmatrix},$$

for some matrices B and C. The result follows from the relation

$$\begin{pmatrix} A_0 & B \\ C & A_1 \end{pmatrix} \begin{pmatrix} Id & -A_0^{-1}B \\ 0 & I \end{pmatrix} = \begin{pmatrix} A_0 & -B \\ C & A_1 - CA_0^{-1}B \end{pmatrix},$$

taking determinants. $\square$

17. General Laplace expansion

First, we need to introduce again some notations. If I (respectively J) is a subset of $\{1, 2, \ldots, m\}$ (respectively a subset of $\{1, 2, \ldots, n\}$) then the notation I' represents the complement of the set I relatively to $\{1, 2, \ldots, m\}$ (respectively, J' represents the complement of J relatively to $\{1, 2, \ldots, n\}$).

The following result formula (2) in Section 2; it was proved by Laplace in 1772.

Let J be a fixed subset of $\{1, 2, \ldots, m\}$. The determinant of a square matrix A of size $m \times m$ satisfies

$$\det A = \sum_{I \in \mathcal{P}_k\{1,2,\ldots,m\}} (-1)^\sigma \det A(I, J) \cdot \det A(I', J'),$$

where $\sigma = \sum_{i \in I} i + \sum_{j \in J} j$.

Proof

Using Eq. 3, it is easy to verify that the left-hand side and the right-hand side contain the same terms, except maybe for signs. Then, it is an easy exercise to verify that these signs are indeed the correct ones.‡

Remark

Formula (2) is the special case $\mathrm{Card}(J) = 1$ of the previous result. Notice also that a similar result holds for I fixed and J running over the set $\mathcal{P}_k\{1, 2, \ldots, m\}$ (proof : use transposition).

18. Sylvester's identity

Consider a matrix $A \in M_n$. Let I be a subset of $\{1, 2, \ldots, n\}$ which contains k elements. Put $b_{ij} = \det\big(A(I \cup \{i\}, I \cup \{j\})\big)$ for and $1 \le i, j \le n$. Then,

$$\det(B) = \big(\det A(I)\big)^{n-k-1} \det(A).$$

Proof

See the book of McDonald‖. □

---

‡ For a more detailed proof see, for example, Maxim Bôcher, *Introduction to higher algebra*, Macmillan, New York, 1907.

‖ B.R. McDonald . — *Linear Algebra over commutative rings*; Marcel Dekker, New York, 1984.

19. Another identity

Once more, we introduce some notations. Let A be a matrix in $M_{m,n}$. Let I be a subset of cardinality k of the set $\{1, 2, \ldots, m\}$. Then the notation $A(I\,;\, j_1, \ldots, j_k)$ represents the matrix whose rows are the rows of A whose indices run over I and whose sth column is column j_s of the matrix $A(I, \{1, 2, \ldots, n\})$.

With the previous notations, we have§,

$$\det A(I\,;\, i_1, \ldots, i_k) \cdot \det A(I\,;\, j_1, \ldots, j_k) = \sum_{t=1}^{k} \det A(I\,;\, i_1, \ldots, i_{s-1}, j_t, i_{s+1}, \ldots, i_k) \times \det A(I\,;\, j_1, \ldots, j_{t-1}, i_s, j_{t+1}, \ldots, j_k).$$

§ G.B. Price . — Some identities in the theory of determinants; *Amer. Math. Monthly*, 54, 1947, p. 75–90.

INDEX OF NAMES

Location of subjects is noted by chapter number: main section number: subsection number.

INDEX

Location of subjects is noted by
chapter number: main section number: subsection number.